Design and Construction of an RFID-enabled Infrastructure

The Next Avatar of the Internet

PUBLISHED TITLES:

Design and Construction of an RFID-enabled Infrastructure: The Next Avatar of the Internet
Nagabhushana Prabhu

Cultural Factors in Systems Design: Decision Making and Action
Robert W. Proctor, Shimon Y. Nof, and Yuehwern Yih

Handbook of Healthcare Delivery Systems
Yuehwern Yih

Manufacturing Productivity in China
Li Zheng, Simin Huang, and Zhihai Zhang

Design and Construction of an RFID-enabled Infrastructure

The Next Avatar of the Internet

Nagabhushana Prabhu

CRC Press
Taylor & Francis Group
Boca Raton London New York

CRC Press is an imprint of the
Taylor & Francis Group, an **informa** business

CRC Press
Taylor & Francis Group
6000 Broken Sound Parkway NW, Suite 300
Boca Raton, FL 33487-2742

First issued in paperback 2019

© 2014 by Taylor & Francis Group, LLC
CRC Press is an imprint of Taylor & Francis Group, an Informa business

No claim to original U.S. Government works

ISBN-13: 978-1-4398-0741-5 (hbk)
ISBN-13: 978-0-367-37918-6 (pbk)

Library of Congress Cataloging-in-Publication Data

Prabhu, Nagabhushana.
  Design and construction of an RFID-enabled infrastructure : the next avatar of the Internet / Nagabhushana Prabhu.
      pages cm. -- (Industrial and systems engineering series)
  Includes bibliographical references and index.
  ISBN 978-1-4398-0741-5 (hbk. : alk. paper)
  1. Radio frequency identification systems. 2. Embedded Internet devices. I. Title.

  TK6570.I34P72 2013
  621.3841'92--dc23   2013034941

Visit the Taylor & Francis Web site at
http://www.taylorandfrancis.com

and the CRC Press Web site at
http://www.crcpress.com

Contents

Preface

The next avatar of the Internet will revolutionize our world. In time, it will provide us a *universal remote control*, enabling us to monitor and control physical objects located anywhere on the planet, using a smart phone. It will make the universe around us *programmable*, allowing us to script the behavior of physical objects with electronic commands. This book is about such an emerging new version of the Internet. Although the technology is still in its embryonic stages, homeowners are already using it to operate their home appliances, such as thermostats and lamps, over the Internet. Not only is the emerging technology enabling us to talk to the objects, it is also enabling objects to talk to us. Plants that need water are already texting their owners to ask for water. Roads and garages are telling drivers in crowded cities where the vacant parking spots are. The physical world has started talking to us and responding to us in unprecedented ways. However, that is just the tip of the iceberg. This book is a narrative about the coming technological tsunami that will transform our relationship with the physical world and usher a disruptive game-changing advance of an unprecedented magnitude.

The emerging infrastructure is called *Internet 2.0*—or *I-2*—in this book. The term warrants elaboration. The current avatar of the Internet, called *Internet 1.0* in the following discussion, is a platform that supports the flow of information traffic across the world. It provides the technology to transport a digital resource, such as a photograph or a digital document, from one computer in the world to another. It is akin to the nervous system in the human body. The nervous system is a labyrinthine network of nerves whose function is to support the flow of information (neuronal signals) and the processing of information. A nervous system by itself, however, cannot sense or actuate the physical world around it. It has to be coupled to sense organs such as eyes and actuators such as muscles for it to gain the capability to see the world around it and to move the objects in the physical world. Internet 1.0, like the bare nervous system, is not endowed with the ability to sense the world around it or actuate the physical objects. It has no "eyes" or "muscles." The emerging technology seeks to enhance Internet 1.0 by endowing it with the capability to sense the physical world around it and actuate the objects. In an anthropomorphic sense I-2 is Internet 1.0 with sense organs and muscles.

The nascent infrastructure was suggestively called the *Internet of Things (IoT)* by previous authors and researchers to highlight that it is intended to interact not only with the cyber resources, such as documents and images, but also with the physical "things" around us. It presented a tantalizing vision of an Internet that interacted with "things," bridging the cyber and physical worlds together.

However, the difference between cyber resources and physical objects, though self-evident from a human perspective, is a contrived distinction from an architectural perspective. From an architectural perspective the differences among a plant sending a message asking for water, a printer sending a message asking for paper, or an application program sending a message asking for input data are rather artificial and rooted, not so much in the inherent distinctions, but in human prejudice. The design philosophy of *universality* that underlies both the Internet and the World Wide Web (see Chapter 8) posits that the core architecture of a global infrastructure must not couple to artificial differences and heterogeneities. Thus, the word "things" in the *Internet of Things*, suggestive of a distinction between cyber and physical resources, is to be avoided in thinking about the new infrastructure. An author's license is invoked in this book to call the infrastructure, Internet 2.0, or I-2, instead. A reader, who prefers the term Internet of Things, may mentally replace *I-2* with *Internet of Things* everywhere in this book. The nuance in terminology, however, is not merely linguistic hairsplitting. The choice of the term *Internet 2.0* underscores the importance of treating physical and nonphysical resources symmetrically from an architectural perspective. It embodies an enduring adherence to universality.

This book is a narrative about the emerging I-2. As with the birthing process of any new paradigm, the emergence of I-2 is also marked by a profusion of ideas, technologies, standards, architectures, projects, initiatives, and proposals. In the ongoing Darwinian struggle, some of the competitors will survive and will find a place in the final I-2 infrastructure. Others will perish. It is not the goal of this book to present a comprehensive survey of all the alternatives that are vying to contribute to the birth of I-2, although, inevitably, some of the alternatives are woven into the fabric of this narrative. Rather, the goal of this book is to tunnel through the profusion of ongoing research and commercial activities to address three questions.

Evangelistic speculations about the potential of I-2 have outpaced the actual progress on its construction. While there has been considerable fragmented organic growth of the infrastructure, the vision of I-2 as a sprawling global infrastructure that supports a seamless interaction among humans, cyber resources, and physical objects remains largely unrealized. At present there are noninteracting islands of activity—Intranets 2.0—in which the vision has been realized to varying extents. However, more than a decade after the vision was articulated we do not even have a globally deployed prototype. The question then is: what are the barriers, if any, to the emergence of I-2 as a global infrastructure? This is a question for which the answer has several dimensions—technical, societal, legal, economic, and political, to name a few. All of the dimensions except the technical dimension are ignored in this narrative. In summary, the first question addressed in this book is: *what are the technical roadblocks for the emergence of I-2?*

The search for an answer to the question posed above threads the narrative in this book through the birthing process and maturation of two of the most successful global infrastructures ever built—the Internet and the World Wide Web (web). The design principles that the architects of the Internet and the web embraced, the environment in which those behemoth infrastructures were conceived and built, and the strategic decisions that were made—both by the architects of the Internet and the web as well as by the federal agencies—contain valuable take-away lessons for architects of any global infrastructure. The comparative histories of the Internet and I-2 uncover some significant differences between the evolution of the Internet on the one hand and that of the I-2 on the other. Those differences are discussed in Chapter 7.

The second question is the following: *what are the essential features that must be incorporated into I-2's architecture?* Several competing architectures have been proposed for I-2. However, no single architecture has emerged as a clear choice. The thinking continues to be fragmented, sometimes unduly distracted by the dichotomy between physical and nonphysical resources. We critically review the architectures of the Internet and the web to identify core guidelines for the architecture of I-2. These guidelines and the work of other researchers, when distilled, suggest that the atomic unit of interaction in I-2 must be a unifying abstraction called *web-enabled service*. Further, the atomic building blocks of the I-2 network must be *service agents*—entities that provision and/or consume web-enabled services—with the dichotomy between physical and nonphysical entities hidden as implementation details inside the service agents. Hiding the dichotomy between physical and nonphysical resources embodies the design principle that such dichotomy is really irrelevant to the architecture of I-2. An attempt to look at the fine-grained structure of a service agent and distinguish between physical and nonphysical resources leads to an unproductive explosion of architectural entropy. A growing number of researchers are recommending service-oriented architecture as the paradigm of choice for I-2. In that sense, the suggestion presented in this book is not new. However, the argument in this book goes beyond recognizing the importance of service-oriented architectures. It consolidates the thinking of previous researchers, even some explicit suggestions by previous researchers to the effect, as well as the lessons learned from the Internet and the web to elevate web-enabled service to be the unifying, all-encompassing, irreducible construct that should be the atom of transaction in I-2. The unifying abstraction is similar to the umbrella construct—the *resource*—in the World Wide Web. The diversity of the digital resources, such as the different file formats, the different types of digital objects (documents, images, videos, application programs), is hidden under the umbrella construct called the "resource," thereby confining the heterogeneity of resources to the edge of the World Wide Web. Doing so enabled the core architecture of the web to remain simple and focus on the flow of a single construct—the resource. Similarly, subsuming physical and nonphysical resources as implementation details inside a *service agent* restricts the bewildering heterogeneity of I-2 to the edge, enabling the core architecture to remain simple.

Finally, the book addresses the question: *how does one build a prototype of I-2?* The growth of both the Internet and the web began with the construction of the respective prototypes. The prototypes not only served as seeds that grew into global infrastructures but, more importantly, provided test beds to vet design alternatives during embryonic stages of growth. Taking a cue from the prototype of the Internet, which exploited the preexisting telephone lines to network the fledgling Internet's nodes, we examine the possibility of using preexisting resources to build a prototype of I-2. The discussion in the book is summarized into a set of technical and strategic recommendations in the final chapter.

Writing about a paradigm that is in its incubation phase poses special challenges. The birthing process of a paradigm involves a close interplay between its underlying architecture and the technologies used to implement the architecture. Maintaining the separation between the architecture on the one hand and the technologies used to implement it on the other is particularly important since our objective is to identify the structural barriers to the evolution of the I-2 paradigm. Structural obstacles cannot be uncovered through an examination of component technologies in depth. An excessively fine-grained discussion of the details of the constituent technologies not only sheds little light on the structural barriers but also has a limited shelf life. On the other hand, an excessively coarse-grained discussion, one that is completely divorced from the details of the constituent technologies, hides the texture of the new paradigm, the details of its birthing pains, and even a clear picture of the structural challenges it is facing. The challenge in writing this book was to sideline the details of the technologies sufficiently to keep the focus on the structural aspects of I-2 while presenting just enough details about the current technologies, standards, and protocols to enable a meaningful discussion of the structural barriers.

The book is organized as follows: The first part of the book sets the emerging I-2 in the context of selected previous game-changing technologies. One of the key pieces in I-2 is the family of bridge technologies that helps connect the cyber and physical worlds. RFID, which is poised to serve as a prominent bridge technology in I-2, is the focus of the second part of the book. In the third part of the book we review the existing global infrastructures, namely, the Internet, the web, and the mobile Internet. We also review selected previous efforts directed at building I-2. The final part of the book is focused on the design of I-2 and construction of its prototype. We start the final part, in Chapter 7, with a comparative review of the evolutions of the Internet and the I-2. The objective of the comparison is to identify notable divergences in their evolutions. In Chapter 8, we consolidate the lessons embodied in the Internet and the web into a set of design guidelines for global infrastructures in general and I-2 in particular. In Chapter 9, we review preliminary notions about services, focusing in particular on a special family of services—the web services—and their variants, the web-enabled services. Against the backdrop of the discussions in the prior chapters we present the architectural guidelines for I-2 in Chapter 10. Chapter 11 is focused on the construction of a prototype for I-2. The discussion in Chapters 7–11 is consolidated, in Chapter 12, into a list of recommendations intended to facilitate the ongoing efforts to build I-2.

I received an enormous amount of support while I was writing this book. It is with gratitude that I acknowledge the assistance I have received. In the early stages of writing I had very helpful conversations on RFID with Sangtae Kim. I have gained a lot of insight into this topic from the conversations I had with Sanjay Sarma. Kevin Ashton, Timothy Berners-Lee (Amy van der Hiel), Vint Cerf and Robert Williams generously granted the permission to use verbatim quotes from their writings; Vint Cerf also permitted the use of a verbatim quote from one of his talks; Stacy Cowley provided the data for Figure 7.2; I am grateful to them for their generosity. I thank Soundar Kumara for reading the entire manuscript and for his valuable comments and feedback. I am, however, fully responsible for the errors that remain in the book. I owe a debt of gratitude to Gavriel Salvendy for getting me started on this project and for his relentless "encouragement" to get me to meet the various deadlines. I am also indebted to Cindy Carelli for all that she has done to accommodate my, often unreasonable, requests for extensions of deadline. It is not an exaggeration to admit that this book would not exist were it not for the support that Gavriel and Cindy extended to me. I thank Amber Donley, Jim McGovern, Galadriel Frond Nair and Kathy Johnson for their patience and their help in converting my draft into a final publishable form. Finally, I would like to thank my wife, kids and parents for their patience and support while this book was being written.

Nagabhushana Prabhu
West Lafayette, Indiana

About the Author

Nagabhushana Prabhu is Reilly Professor of Industrial Engineering at Purdue University. He holds a B.Tech. in computer science and engineering from the Indian Institute of Technology, Bombay, a Ph.D. in computer science from the Courant Institute of Mathematical Sciences, New York University and a Ph.D. in theoretical physics from the Massachusetts Institute of Technology. His other interests include the theory of optimization, computational pathology and quantum field theory.

Section I

Introduction

1

Game-Changing Technologies

The year was 1854. At the Crystal Palace Exposition in New York a curious crowd had gathered around a stunt that was in progress. A man stood on a wooden platform that was suspended by a rope high above the ground. (See Figure 1.1.) Two vertical guard rails flanked the platform on either side. As the crowd watched, an axe-man severed the rope sending the platform into what seemed certain to be a free fall. However, instead of going into a free fall, to the crowd's surprise, the platform fell only a few inches before a braking apparatus, attached to the platform, latched into the vertical guard rails arresting the platform's fall. In this staged stunt, the man on the platform, Elisha Graves Otis, demonstrated that his invention, the *braking apparatus for elevators*, could ensure the safety of the passengers even if the elevator cable were to snap [Srinivasan 2005].

FIGURE 1.1
Elisha Otis demonstrating his invention, the safety elevator. (Courtesy Corbis Images.)

It is unclear if the bemused crowd, or even Otis himself, fully appreciated the transformative impact his invention was poised to have.[*] His braking apparatus ushered the age of safety elevators. Initially his invention was used to meet the obvious and preexisting needs—transportation up and down tall buildings and steep hillsides. The first passenger elevator was installed in New York in 1857 [*Elevator World* 2012]. By 1873 more than 2000 passenger elevators were operating all over America, in buildings that were just a few stories high [Korom 2008]. The safety elevators were also being used to transport coal and lumber up and down the steep hillside slopes [*New World Encyclopedia* 2012]. But these were only the first wave of applications.

The greater impact of the safety elevators came, not in the first wave of applications, which focused on preexisting needs, but instead in the second wave of applications that were about new, unanticipated possibilities suggested by the invention itself. The invention of the safety elevator initiated a spectacular vertical growth of metropolises. Prior to Otis' invention, buildings were only a few stories tall and urban population densities relatively small. The safety elevator led to the construction of buildings and skyscrapers of unprecedented height. The emerging skyscrapers, in turn, drove advances in materials engineering and architectural design. The vertical growth of cities led to a rapid rise in the population densities giving rise to new urban ecosystems. The seemingly simple invention that the crowd witnessed in 1854 turned out to be a game-changing advance that irreversibly transformed the urban landscape.

Game-Changing Inventions

The safety elevator was neither the first nor necessarily the most prominent game-changing invention. The history of technological progress is punctuated with game-changing advances many of which have unobtrusively integrated themselves into the fabric of modern life. In fact, Mark Weiser argues that their very ability to blend themselves into our everyday lives is a defining hallmark of great inventions [Weiser 1991]. Several game-changing technologies meet Weiser's criterion and have become an inextricable part of our everyday lives, as the following examples show.

Regarded the most important invention of the second millennium [Mainz 1997], the *movable type printing press*, invented by Johannes Gutenberg, around 1440 A.D., was a profound advance that has become an inextricable part of everyday life, the pivotal role it played in advancing human civilization buried under layers of history. The printing method that was used in the east, prior to Gutenberg's invention, could produce about 40 copies per

---

[*] I thank Sangtae Kim for articulating the significance of Elisha Otis' invention.

day [Hye-Bong 1993]. By 1600 the Gutenberg press was producing about 3600 copies a day [Wolf 1974] representing a 90-fold increase in the speed of replication. It ushered the era of mass communication, facilitated the Renaissance, and led to the scientific revolution. As much as any other invention it promoted the dissemination and cross-fertilization of human thought.

Another game-changing advance that also facilitated interaction among people was the invention of a powered heavier-than-air flying machine, the *airplane*. Prior to the advent of aviation technology travel by land or by sea was an order of magnitude slower, which in turn made essential services, such as the postal service, much slower than it is today. By transforming travel and logistics services aviation technology has become an essential, and often invisible, part of modern life.

The invention of the *semiconductor transistor* in 1947, by John Bardeen, Walter Brattain, and William Shockley, is regarded by some as one of 20th century's greatest inventions [Price 2004]. It enabled a dramatic increase in the density of digital circuitry and ushered the modern electronic age. Modern electronic devices, from laptop computers and implantable pacemakers to artificial satellites, would be inconceivable without the miniaturization that was made possible by semiconductor transistors.

Artificial satellites themselves were a game-changing invention. From wireless transmission of voice and text data to weather prediction and *GPS*-based navigation, satellites are participating invisibly in several aspects of our daily lives. The vast transcontinental bandwidth for text, voice, and video data that we have become accustomed to was infeasible in the pre-satellite world.

Another spin-off of the semiconductor technology was the invention of the *charge-coupled device (CCD) technology*, which gave birth to digital photography [Aaland and Burger 1992]. It was a disruptive advance over the film-based photography that preceded it. It has enabled us to record events with unprecedented ease and agility.

A third game-changing spin-off of the semiconductor technology is the *mobile phone*. The first handheld phone was prototyped in the early 1970s by Martin Cooper. While there were very few mobile subscriptions prior to 1990 [Ferris 2010] today there are about 78.6 mobile subscription lines for every 100 people in the world. Figure 1.2 shows the dramatic increase in the mobile subscriptions over the past two decades. By Weiser's criterion, mobile communication devices indeed represent a profound technology in that they have become indistinguishable from the fabric of everyday life. In fact, mobile communication devices have penetrated everyday life to a greater extent than another prominent technology that originated alongside mobile telephony—*personal computing*.

Like mobile telephones personal computers were also developed in the early 1970s. The advent of personal computers marked the dawn of the information age. Their widespread adoption, and their pervasion into everyday life was made possible to no small extent by another often under-appreciated invention—the *computer mouse*.

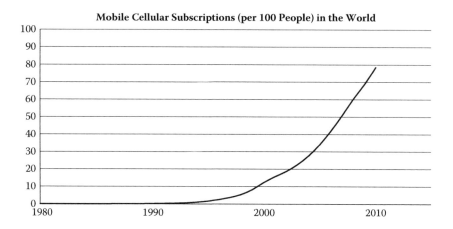

FIGURE 1.2
The data in this figure are taken from UNdata [World Bank 2011b].

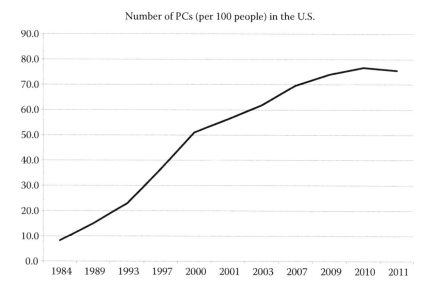

FIGURE 1.3
The data shown in this figure is taken from UNdata [US Census 2011].

   The computer mouse, a critical enabler of graphical interaction between a human and a computer, was invented by Douglas Engelbart in 1964 [Engelbart 2000]. The mouse enabled a more intuitive visual interaction between humans and computers, paving the way for a broader adoption of personal computers in everyday life. Besides the mouse the proliferation of personal computers was also facilitated by the development of user-friendly operating systems. Figure 1.3 illustrates the growth of the personal computer's usage in the United States.

User-friendly personal computers enabled widespread diffusion of computing technology into households, in turn, setting the stage for the emergence of two global information infrastructures of unprecedented size and reach—the *Internet* and the *World Wide Web*.

The Internet and the World Wide Web are arguably two of the most transformative game-changing advances in human history. They have touched practically every aspect of human life and in the true tradition of profound technologies have inseparably intertwined themselves into the fabric of everyday life.

The Internet and the World Wide Web

The Internet and the World Wide Web (web) are often conflated. The distinction between them is brought into sharper focus in the simplified illustration of the two infrastructures shown in Figure 1.4.

Crudely, the Internet is a network of hardware devices interconnected by communication links as shown in the bottom plane in Figure 1.4. The communication links support data transfer among the devices. The devices are classified into two groups—*routers* and *end nodes*. Data flowing through the

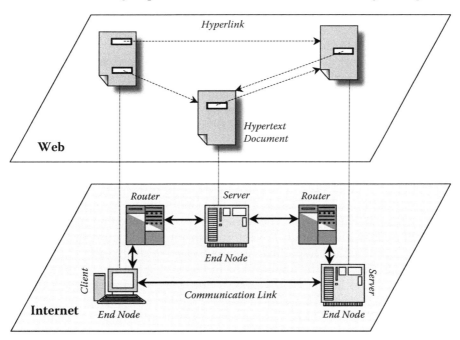

FIGURE 1.4
A simplified illustration of the Internet, the web, and the relation between them.

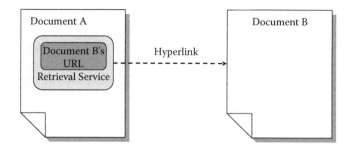

FIGURE 1.5
Anatomy of a hyperlink.

Internet are generated and consumed at the end nodes. Routers are intermediate hardware devices that do not generate or consume data but instead merely facilitate the flow of data among the end nodes.

The end nodes of the Internet host electronic resources, such as hypertex documents, images, video, and music files. The resources are shown in the upper plane of Figure 1.4, with vertical lines linking the resources to the hardware units on which they reside. Prior to the development of the *World Wide Web* (web), a resource at an end node was unaware of the existence of other resources on the Internet. That is, the hyperlink pointers shown in the upper plane did not exist in the pre-web Internet. The web, a framework created by Berners-Lee (Berners-Lee 1989), brought this passive collection of resources to life through two enhancements.

First, the web provided a scheme to assign a globally unique address to every resource, making resources on the Internet addressable. The unique address—called a *Uniform Resource Locator (URL)*—specifies the location of the resource on the Internet. Thus, one can point to Document B from within Document A by inserting Document B's URL in Document A, as shown in Figure 1.5.

The URL by itself, however, is a passive reference. Berners-Lee's seminal contribution was that he realized a passive reference—such as a URL—can be bundled with a retrieval service by exploiting the data transport capability of the Internet. A passive URL bundled with such a retrieval service becomes an active *hyperlink*. The retrieval service, invoked by clicking on the hyperlink, tunnels through the Internet to retrieve Document B. A document with embedded hyperlinks is called a *hypertext document*.

The web, shown in the upper plane in Figure 1.4, is a library of hypertext documents. The library of documents and the hyperlink pointers among them form a network—a web—that was intended to create, in Berners-Lee's words [Berners-Lee 1989], "a universal linked information system." The retrieval service provided by the hyperlinks enables users of the web to navigate the web of documents easily using the intuitive user-friendly operation of clicking on hyperlinks.

From the perspective of the web, there are two types of end nodes in the Internet—labeled *clients* and *servers* in Figure 1.4. In order for a document to be visible on the web it must be hosted by a web server (or just server). The end nodes of the Internet that do not function as servers are called clients from the web's perspective. While documents on server nodes are address-able on the web, documents on client nodes are not. Consequently, hyper-links can point to documents on servers, but not to documents on clients (see Figure 1.4). The Internet is discussed in greater detail in Chapter 3 and the web in Chapter 4.

Applications and Size of the Internet and the Web

The Internet originated as loosely coupled networks of computers in the late 1960s and the early 1970s. The first wave of applications of the Internet focused, not surprisingly, on a preexisting need—transfer of files among computers. Prior to the birth of the Internet, files were being transported physically using storage media such as magnetic tapes, which made the transfer process rather slow. The Internet made it possible to transfer files over communication links, in real time; the popular *File Transfer Protocol* (FTP) was developed in 1971 to enable file transfers using standard system-independent commands [Bhushan 1971]. It set the stage for the second wave of functionalities, which was heralded by the development of a killer appli-cation by Ray Tomlinson—the *electronic mail* or *email*.

The development of email was a transformative event in human history. It dramatically lowered the barrier for person-to-person real-time written communication. As much as any other advance, the email has helped knit the world population into a networked community.

The second wave of applications also witnessed the emergence of online social communities—the *newsgroups*. The newsgroups were user-driven global electronic bulletin boards open to everyone connected to the Internet. The emergence of such ad hoc topic-centered newsgroups, open to world-wide participation, was a new paradigm that was inconceivable in the pre-Internet world.

Although email and newsgroups made unprecedented strides in network-ing the world community, their impact was dwarfed by that of the most suc-cessful application to ever run on the Internet—the World Wide Web. The rapid increase in the Internet usage, shown in Figure 1.6, began with the development of the web by Berners-Lee in 1989–91 [Berners-Lee 1996].

The first wave of applications of the web indeed centered on creating a universal linked information system envisioned in Berners-Lee's original proposal [Berners-Lee 1989] and saw the growth of a "docuverse"—a uni-verse of mostly static interlinked documents. End users of the web were

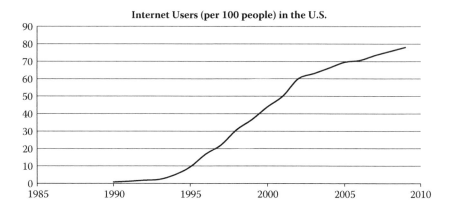

FIGURE 1.6
The data shown in this figure are taken from [World Bank 2011a].

mostly consumers of the information that was published on the web servers of larger organizations. In other words, the first wave of applications
involved hierarchical dissemination of information rather than peer-to-peer
information exchange.

The significant impact of the web, however, came in the second wave of
applications, which made the web more than just a static docuverse. Activities
such as shopping, banking, and bill payments, which were being done in
person, are increasingly being done online on the web. The web has given
birth to social media and social networking, empowering individuals to be
not only consumers of content but also producers of content. Previously, the
dissemination of information was controlled by governments and news agencies. Social media has created parallel user-driven communication channels
that have already had significant geopolitical impact [Vargas 2012], played an
invaluable role in the aftermath of natural disasters [Wallop 2011], and appear
poised to shape modern life in disruptive ways in the years ahead [Rainie and
Wellman 2012]. In the tradition of profound advances the Internet and the
web have irreversibly woven themselves into our everyday lives.

The growth of the Internet, illustrated in Figure 1.6, raises the question
about the size of the web. While it is difficult to estimate the size of the web
precisely Google's engineers Alpert and Hajaj reported in 2008 that they had
tracked over a trillion different Uniform Resource Locators on the web and
that the number of web pages was increasing at a rate of about a few billion every day [Alpert and Hajaj 2008]. That is, the web has more than 140
URLs for every person on the planet. At the end of 2011 more than two out
of every seven people on the planet were connected to the Internet [Internet
Stats 2011]. These statistics provide a quantitative indication of the pervasive
global impact that the Internet and the web have had.

An Emerging Game-Changing Technology

The Internet and the web (including the mobile Internet and the mobile web, discussed in Chapter 5) make up the current *cyber infrastructure*. Despite its spectacular growth and pervasive impact, stripped of all the bells and whistles, the cyber infrastructure has a rather limited functionality. It merely transports digital data between two devices connected to it. Admittedly, the capability to transmit data across the globe in real time has had an unprecedented impact on several aspects of human life, and has ushered us into the information age. However, at present, the cyber infrastructure is still just a labyrinth of nodes connected by communication highways with a data transportation service operating on it.

Although we have entered the information age physical objects continue to play an important role in modern life. For example, we still use objects such as lamps to illuminate our spaces, refrigerators to store food, and automobiles to transport us. Every day we interact with objects such as clothes, shoes, furniture and other personal effects. It is estimated that a typical person, living in an urban setting has 1000 to 5000 physical objects in his/her surroundings [Waldner 2010]. In addition, we use large physical infrastructures such as buildings, garages, roads, and bridges.

One of the biggest shortcomings of the current cyber infrastructure is that it does not have the capability to interact with the world of physical objects without human assistance. For example, objects purchased at a store have to be scanned by a human at the checkout counter before the data about the purchase enters the cyber infrastructure (as an update to the inventory database). A misplaced book in a library must still be manually searched, and its status in the database must be updated by a human. The current cyber infrastructure, by itself, cannot sense the location of a book in the library. If a hazardous situation, such as a leak from a burst pipeline, develops on city streets the current cyber infrastructure cannot sense the hazard and automatically reprogram the traffic lights to divert traffic away. Staff in most hospitals still manually search for misplaced medical equipment instead of making the cyber infrastructure interrogate the physical surroundings to determine the real-time locations of the instruments. The cyber infrastructure is not being used in all airports to maintain real-time awareness of the locations of the passengers' checked bags. According to the 2010 Baggage Report more than 800,000 passenger bags were lost by airlines in 2009 [SITA 2010]. Lost bags are still being tracked by humans. As these examples show, the current cyber infrastructure is mostly "blind." It cannot see much of the physical world without human help. And it has no "limbs." For the most part it cannot make objects in the physical world do things.

In a sense the current cyber infrastructure—comprising the Internet and all the devices connected to it—is like the human nervous system. The nervous system can process information and transport signals across the network of nerves. However, unless it is coupled to sense organs like eyes and actuators like muscles the nervous system, by itself, cannot interact with the physical world. It cannot obtain information about the physical world without the assistance of sense organs. Nor can it, on its own, move physical objects. Similarly, while the current cyber infrastructure can process information and move it around most of the information in the infrastructure has to be put into it by humans. In anthropomorphic terms, the current cyber infrastructure is like an embryo whose nervous system is well developed and whose sensory and motor capabilities are yet to develop.

Endowing the cyber infrastructure with the capability to sense the physical objects and to actuate objects is the objective of an emerging technology that we call **Internet 2.0**, abbreviated to **I-2**.

I-2 is envisioned to be an enhancement of the current cyber infrastructure and a natural extension of it. It seeks to digitally enhance objects in the physical world to make them visible to the cyber infrastructure. Using the current cyber infrastructure as an intermediary, I-2 seeks to enable physical objects to interact and communicate with each other and make decisions without human intervention. The grand vision of I-2 is that it will integrate the cyber and physical worlds into a single giant seamless infrastructure that will function as a primitive globally distributed "organism," that can sense its surroundings, process sensory data, and actuate responses, without human intervention.

For example, I-2 envisions a future in which a home automation system would acquire the real-time Doppler radar data from the web, foresee imminent rain, and turn off the lawn sprinkler system, without human intervention. Parking spots in garages and on the streets would proactively communicate their real-time status, enabling drivers to see the parking spots available in their vicinity on their handheld displays. Hospital staff would be able to instantaneously locate misplaced medical equipment, such as defibrillators and infusion pumps, using a web browser. Surgical consumables and instruments such as sponges and scalpels, inadvertently left behind inside patients during surgery, would remind the surgeons to retrieve forgotten items before the surgical incision is closed. Buildings and bridges would communicate data about their structural health, making it possible to anticipate and preempt catastrophic fractures. A plant would email its owner when it needs water. And so the list goes on.

The above evangelistic description of the future is technologically feasible. Various groups around the world are already engaged in the task of building the new infrastructure. However, the progress has been modest to date despite considerable worldwide efforts to build and deploy the I-2 infrastructure. In the following chapters, we take a closer look at the issues involved in building I-2, culminating in a discussion of the architectural imperatives and a roadmap for constructing a prototype. In the remainder of this chapter we continue with a coarse-grained description of the anatomy of the I-2 infrastructure, which will be brought into sharper focus in later chapters.

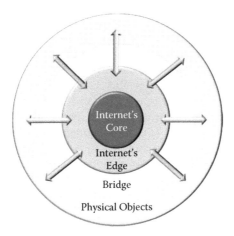

FIGURE 1.7
Illustration of the hardware framework of I-2.

Bridging the Cyber and Physical Worlds

The task of building the new I-2 infrastructure can be broken down into five subtasks. Physical objects must be digitally enhanced to make them visible to the cyber infrastructure and, where appropriate, communicate with the cyber infrastructure and with each other. Second, bridge technologies need to be deployed to connect the physical and cyber worlds. Third, the core of the current Internet needs to be enhanced to handle the dramatic increase in data traffic that would result once the cyber and physical worlds are bridged. Fourth, the edge of the Internet needs to be enhanced to endow it with the capability to communicate with the physical world. These hardware enhancements are illustrated in Figure 1.7. Finally, the I-2 infrastructure must be based on a new architecture that folds the vast heterogeneity of physical and cyber resources into a simple uniform framework enabling interoperability among the physical as well as the nonphysical resources. We elaborate on these subtasks briefly below, deferring a more detailed discussion to the subsequent chapters.

Digital Enhancement of Physical Objects

The physical world comprises objects with which we interact, encompassing everything from coffee mugs to airplanes. A common object, such as an ordinary table lamp, cannot communicate with the Internet. However, by embedding the necessary intelligence in a table lamp it can be given an identity on the Internet, making it possible to operate the lamp from across the world. A book in a library can be affixed with an *RFID (Radio Frequency IDentification)* tag, making it possible for an enhanced end node of the Internet, say, a computer equipped with an RFID reader, to detect the book wirelessly. Such digital enhancement of physical objects—ranging from affixing RFID tags on them to housing intelligence inside them—enables electronic communication between the enhanced physical objects and the cyber infrastructure, without human involvement.

The Bridge

The Internet comprises two parts as shown in Figure 1.7: (1) the part that is invisible to end users, which comprises the large backbone networks that carry the world's Internet traffic together with the standards and software that govern the networks—collectively called the *core,* and (2) the part that is closer to an end user—such as end users' computers and mobile devices—collectively called the *edge.* The precise boundary between core and edge is not important. In I-2 the edge of the Internet and the space of physical objects are to be connected by a set of bridge technologies—collectively called the *bridge.*

The *bridge*—a set of (mostly wireless) communication technologies that link the Internet and the physical world—is the key hardware component of I-2. A prominent example of such a wireless communication technology is the *Radio Frequency IDentification (RFID)* technology, which is the subject of the next chapter.

An RFID bridge comprises an RFID tag affixed to a physical object and an RFID reader connected to an end node of the Internet, as shown in Figure 1.8. A tag and a reader communicate over a wireless link. An RFID reader can wirelessly sense a tag and retrieve information stored on it using which the reader can obtain information about the object to which the tag is affixed. Thus, an RFID reader serves as an "eye" for the Internet, making the Internet capable of "seeing" digitally enhanced objects. It also enables the cyber infrastructure to send commands to actuate physical objects.

At present such technologies are being deployed to create *intranets of things,* small ecosystems in which the cyber and physical worlds are connected [Sundmaeker et al. 2010]. An example of such an intranet is the RFID-enabled baggage handling system that is being used in a few airports [Wessel 2009]. Tagging passengers' bags with RFID transponders enables the airports to automate the process of transferring passenger bags between connecting

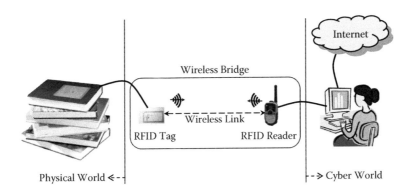

FIGURE 1.8
A wireless bridge between the physical and cyber worlds.

flights, thereby speeding up the process and decreasing the errors in routing bags. RFID-enabled baggage handling is discussed in greater detail in the next chapter. Another example is the RFID-enabled inventory management system used to track products in stores [Roberti 2010a]. These represent applications in which the cyber and physical worlds are bridged within an organization—an airport and a retail store in the above examples. Although the currently operational infrastructures that bridge the cyber and physical worlds are intranets and not the global I-2, they demonstrate the feasibility of bridging the cyber and physical worlds using wireless technologies such as RFID.

The Core

Every object connected to the I-2 infrastructure is expected to have a globally unique identity. IPv4, the addressing scheme that is still being used on the Internet, can assign unique addresses to about 4.3 billion entities. In contrast, estimates of the number of things that will be connected to the Internet by 2020 range from 50 billion [Vestberg 2010] to over a trillion [IBM 2010], making IPv4 vastly inadequate to support the predicted size of I-2. IPv4 is being phased out and is being replaced by the 128-bit IPv6 addressing scheme, which provides enough addresses to uniquely label every object in the foreseeable future.

Connecting billions of devices and physical objects to I-2 presents other challenges besides addressability. Each device that is capable of generating data adds to the data traffic load on the Internet. The short-term forecast for the amount of data flowing through the Internet is shown in Figure 1.9. The data are expressed in *exabytes* (about a million terabytes) [Cisco 2011a]. The

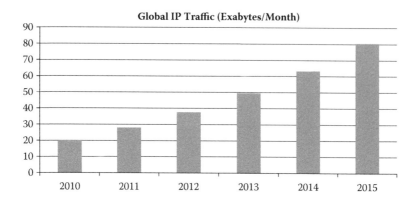

FIGURE 1.9
Forecasted data traffic on the Internet [CISCO 2011a].

estimates predict a fourfold increase in the Internet traffic over just a 5-year horizon. The more notable statistic is that by the end of 2011, the number of networked devices is expected to have exceeded the number of people on earth. By 2015 the number of networked devices is expected to be more than twice the number of people on earth [Cisco 2011b]. That is, the number of networked devices is expected to double in about 4 years. If such doubling continues, facilitated by the growth of I-2, and each such device adds to the Internet traffic, then the traffic can be expected to grow exponentially in the near future. Unless the Internet infrastructure is enhanced to handle the rising traffic it could become vulnerable to implosive events such as *congestive collapse*, a phenomenon that was predicted by John Nagle in 1984 [Nagle 1984] and first observed in October 1986 [Mankin 1991].

A congestive collapse is an abrupt degradation in the bandwidth of the Internet, often by more than an order of magnitude, caused by congestion-induced positive feedback instability [Jacobson and Karels 1988; Jacobson 2009]. Specifically, if the Internet attempts to route data through routers at a higher rate than the routers can handle, then they discard some of the incoming data. Upon sensing the loss of the data the sender would then attempt to re-transmit additional copies of the lost data exacerbating the problem at the congested nodes and leading to positive feedback instability. In the congestive collapse that occurred in October 1986, the bandwidth the NSFNET's phase-I backbone was observed to drop about three orders of magnitude from 32 kbps to 40 bps [ISOC 2010]. Although congestion control protocols [Floyd 2000] were implemented following the 1986 collapse, the Internet had another congestive collapse, as recently as December 2006, following the Taiwan earthquake that damaged undersea Internet cables [Raghavan et al. 2008]. With the predicted fourfold increase in the data traffic during 2010–15, and the flood of additional traffic that will burden the Internet as I-2 grows, the Internet backbone's capacity needs to be enhanced to make it capable of handling the expected growth in the data traffic.

The Edge

Even if the Internet is enhanced to support the data traffic among the billions of objects that will connect to I-2 one is still left with the challenge of ensuring *interoperability* among the objects. Objects that will connect to I-2 will be characterized by vast heterogeneity. Objects small and large—from key chains to airplanes—as well as living and nonliving things—such as cattle and briefcases—are expected to acquire identities in I-2. They will have varied features and functionalities. I-2 seeks to weave all such heterogeneous objects into one large seamless ecosystem in which they can interoperate with one another without human intervention. Such a grand vision of interoperability among heterogeneous devices harks back to a similar vision that guided the architects of the Internet as they set about internetworking heterogeneous local networks into a global infrastructure in which disparate networks could interoperate. The Internet architecture addresses the interoperability issue by pushing the intelligence needed for interoperability to the edge of the network—the devices that connect to the Internet. Similar standards and architectural features need to be deployed at the edge of the I-2 to support interactions among objects. The necessary architectural features are discussed in greater length in Chapters 10 and 11.

Summary

The history of technological progress is punctuated by game-changing inventions that have induced disruptive advances. Whereas the first wave of applications of such advances focused on solving preexisting problems, their larger impact has been in the second wave of applications, which addressed not the preexisting needs but rather new unanticipated possibilities suggested by the inventions themselves.

In the tradition of previous game-changing advances, the emerging I-2 infrastructure appears poised to have a disruptive impact on the modern world. I-2 envisions an unprecedented integration of the cyber and physical worlds that enables the hitherto voiceless physical objects to communicate with the cyber infrastructure and with each other, without human intervention. While the first wave of applications of I-2 can be expected to use the technology to address preexisting needs, the second wave of applications will likely exploit the increased autonomy in machine-to-machine interactions and transform modern life in profound ways.

Currently, the cyber and physical worlds have been bridged in relatively small, island infrastructures that are largely disconnected from one another. A global I-2 infrastructure, akin to the Internet and representing a planetwide integration of the cyber and physical worlds, is yet to emerge. Building the global I-2 infrastructure requires digital enhancement of physical objects, systemic enhancements of the Internet's core and edge, pervasive deployment of bridge technologies, and a new architecture that is custom-designed for I-2. The technologies and paradigms relevant to I-2, the challenges facing it, the architectural imperatives for I-2, and a roadmap for building a prototype are discussed in greater detail in the following chapters.

Section II

Wireless Bridge

Overview

The world of physical objects and the cyber infrastructure are already connected by wireless bridges in several operational systems. For example, the surveillance cameras in electronic toll booths, used to retrieve the license plate numbers of vehicles, act as an optical bridge between the physical objects (the automobiles) and the cyber infrastructure (the toll booth's computing system). The optical bridge relies on the visible band of the electromagnetic spectrum. In addition to the visible band, other bands of the electromagnetic spectrum, such as infrared waves and radio waves, are also being used to bridge the cyber and physical worlds. Infrared sensors, for example, are being used to detect objects in the near field. Radars, operating in the radio band of the electromagnetic spectrum, are being used to sense aircrafts. Technologies such as RFID (Radio Frequency IDentification) and NFC (Near Field Communication) are also exploiting the radio wave band of the electromagnetic spectrum for wireless communication between physical objects and the cyber infrastructure.

Besides electromagnetic waves, other waves or disturbances are also being exploited to bridge the cyber and physical worlds. For example, seismic sensors enable the cyber infrastructure to "see" seismic disturbances. Underwater acoustic sensors are being used to feed aquatic pollution data to the cyber infrastructure [Akyildiz et al. 2005]. Pressure sensors deployed on ocean floors are being used to detect pressure changes associated with tsunamis and communicate data about tsunami events to the cyber infrastructures through satellite links [DART 2012].

Thus, bridging the cyber and physical worlds is not a new notion. What is lacking, however, is an Internet-scale integration of the cyber and physical worlds, and a substrate framework that facilitates autonomous interoperability among the cyber and the physical resources.

The aforementioned examples fall within the purview of what has been called the "bridge" in Figure 1.7. As the examples show, several types of technologies are expected to constitute the bridge for the I-2 infrastructure. In Chapter 2 we discuss the details of one such bridge technology—the RFID—to bring the issues and challenges associated with bridge technologies into sharper focus within a concrete context. The discussion of I-2 in the later chapters is independent of the specific details of the bridge technology.

2

RFID Technology and Embedded Intelligence

Radio Frequency IDentification (RFID) technology originated around the second World War [Roberti 2004], driven by a need to distinguish between hostile and friendly airplanes. While a radar could detect incoming airplanes it could not distinguish between enemy airplanes and returning friendly airplanes. To solve this problem the British military developed an Identify Friend or Foe (IFF) system by attaching a transponder to the British aircrafts. The radio waves emitted by a radar detector woke up the transponder, which transmitted a radio signal back to the base in response to help the base identify it as a friendly airplane. The current RFID technology is modeled after the IFF system.

In modern RFID technology, the transmitter and receiver are merged into a single transceiver called the RFID *reader*, or just "a reader" for short. While the IFF system was used to distinguish between just two categories—friend and foe—modern applications, such as item-level tagging in retail stores, seek to distinguish among a much larger number of categories of objects. A typical modern RFID tag stores a serial number that is about 96 bits long. When interrogated by a reader, a tag transmits its serial number enabling the reader to distinguish among about 10^{28} different tags. Figure 2.1 is a schematic illustration of how the RFID tags are used today. Figure 2.2 is a picture of an actual RFID tag.

Attaching a tag to an object is in some ways akin to putting a license plate on a vehicle. The serial number stored on a tag, like the numbers and characters on a license plate, uniquely identifies the object to which the tag is attached. Like the information on a license plate, the serial number itself does not contain much useful information about the object. The information on a license plate, however, can be used to retrieve information about the details about the vehicle by accessing the associated database. Similarly, the serial number on a tag can be used to retrieve information about the object to which the tag is attached, by interrogating an associated database. The data flow is illustrated in Figure 2.3. In the next section we take a closer look at a modern RFID tag, also called an RFID transponder.

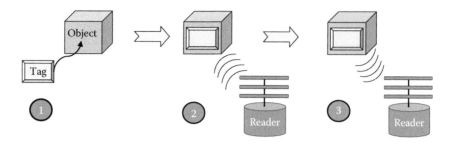

FIGURE 2.1
Schematic illustration of the RFID technology. ① The RFID tag is attached to the object to be tracked, making the object visible (in radio frequency) to RFID readers. ② A reader, which seeks to track the object, emits radio-frequency waves that are picked up by the RFID tag. ③ In response, the RFID tag transmits a radio-frequency signal containing the digital identifier stored on the tag.

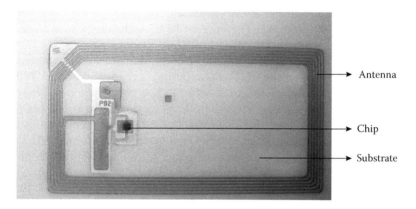

FIGURE 2.2
An actual RFID tag (courtesy Shutterstock). The tag comprises electronic circuitry on a chip and an antenna that is used to transmit the digital information stored on the chip. The dimensions of an RFID tag range from a fraction of a millimeter to a few centimeters.

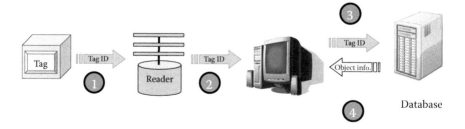

FIGURE 2.3
Data flow in RFID. ① When interrogated by a reader, the tag transmits its ID to the reader. ② The reader in turn routes the ID to a computer connected to the cyber infrastructure. ③ The computer uses the ID as a key to ④ retrieve the information about the tagged object from a database.

Anatomy of an RFID Transponder

A basic RFID transponder, also called RFID tag, has three components: a *chip*, an *antenna*, and a *substrate* in which the chip and antenna are embedded. (See Figure 2.2.) Additionally, the tag could also house a battery that powers the chip and sensors such as temperature or humidity sensors. Tags that have sensors are called *sensor tags*. A detailed overview of the various kinds of RFID tags and their operating principles can be found in Finkenzeller [2010].

The antenna in a tag electromagnetically couples the tag to an interrogating RFID reader. In passive tags the antenna funnels the energy in the radio waves coming from a reader to power the circuitry in the tag's chip. The antenna also enables the tag to broadcast the return communication to the reader.

The chip on a tag contains, among other data, a number that acts as a unique identifier of the tag. For example, an EPCglobal[*] Class 1 Gen 2 RFID tag stores the data in four separate memory banks.

> **Bank 0:** This bank stores a *kill password* and an *access password*. These passwords are used for authentication. To kill a tag a reader needs to provide the kill password stored on the tag. Similarly, to gain privileged access to the contents of the tag a reader needs to provide the tag the access password.
>
> **Bank 1:** This bank stores two control fields—CRC and PC—and the EPC, a globally unique identifier for the object to which the tag is attached. The CRC (Cyclic Redundancy Check) field stores a number that is used to verify the integrity of the data retrieved from the chip. The PC (Protocol Control) field contains information both about the data layout on the chip—such as whether Bank 3 contains any data—as well as details about the protocol used to encode the data on the chip (whether the data conforms to the EPCglobal standard, and if it does not, information about the standard used to encode the data). The EPC (Electronic Product Code), arguably the most important data on the tag, is an identifier that is unique to the tag. The format of the EPC is described below (see Figure 2.5).
>
> **Bank 2:** This bank contains information about the tag itself, including details about the manufacturer of the tag.
>
> **Bank 3:** This bank is earmarked for holding user specified data.

The bit level details of the data format in the four banks described above are elaborated in [GS1 EPC Tag Data Standard 1.6 2011]. In addition to storing the EPC and other data, the chip also houses the circuitry needed to implement

---

[*] EPCglobal is an organization that is working to promote the standardization of RFID technology.

FIGURE 2.4
The 12-digit Universal Product Code used in barcode labels (courtesy upccode.net).

such functions as running anticollision algorithms (described later). If the tag is coupled to external sensors, then the chip would house the circuitry needed to transmit the sensor data to the reader.

Electronic Product Code

An RFID tag is often compared to a barcode label. A barcode label has a series of vertical white and black lines as shown in Figure 2.4. The barcode labels use the *Universal Product Code* (UPC), which is a system of 12-digit numbers used to identify object classes. The first (leftmost) digit in a UPC specifies what the remaining digits represent.[*] Digits 2–6 represent the manufacturer ID and digits 7–11 represent the object class. The last digit is used for error checking[†] to guard against errors in either scanning or entering the digits. The UPC contains only a manufacturer ID and an object class but does not contain a serial number of the specific instance of the object. For example, two laptops based on the same model and made by the same manufacturer would get the same UPC label, although the serial numbers on the laptops would be different.

In contrast to the UPC, the *Electronic Product Code* (EPC) used in RFID tags allow serial numbers of the objects to be included in the label. The structure of a 96-bit EPC label is shown in Figure 2.5. The number is represented in hexadecimal code.[‡] (See [Banks et al. 2007])

If EPC is used to label the boxes containing the two laptops in the previous example, then the first 60 bits of the EPC labels would be identical. Bits 60-95,

---

[*] For example, if the first digit is 0 then the next ten digits represent a regular UPC code. On the other hand, if the first digit is a 5, then the next 10 digits represent a coupon code. The 12-digit numbering schemes, shown here, is the so-called UFC-A scheme (Obal 2004).

[†] The last digit is computed as follows: add to three times the sum of the odd-numbered digits, all the even numbered digits and take the remainder obtained by dividing the resulting number by 10. If the remainder is nonzero, then the last digit is the number that must be added to the remainder to get 10. For the above example the calculation yields 3 * 25 + 20 = 95. The remainder obtained by dividing 95 by 10 is 5, to which one needs to add 5 to get 10. So the last digit is 5. (See [GS1 2013])

[‡] In hexadecimal code a group of 4 bits is used to represent a single hexadecimal digit with, A = 10, B = 11, C = 12, D = 13, E = 14, and F = 15.

05.A98EDF0.2055AF.1005B2710

Header	Manager	Object Class	Unique Identifier
Bits 0–7	Bits 8–35	Bits 36–59	Bits 60–95

FIGURE 2.5
The format of a 96-bit EPC identifier. Besides the manager and object class information, it also contains a field for the serial number of the object. The number shown above was generated at random for the purpose of illustration.

however, would contain different numbers corresponding to identifiers that are unique to the two laptops.

EPC earmarks fields for a manager (manufacturer) and an object class since it is geared toward supply chain applications. On the other hand the uCode naming scheme, discussed in Chapter 6, is an alternative to EPC that enables assignment of identifiers to more general objects and even abstract notions.

The capability to assign unique labels to different objects of the same object class and manufacturer, made possible by the EPC scheme or the uCode scheme, is not the only advantage that RFID tags have over barcode labels. Reading a barcode label requires the label to be in the line of sight of the reader. The RFID readers, on the other hand, can read tags that are not in the line of sight as long as the tags are within the range of the reader. This seemingly simple difference is significant in bridging the cyber and physical worlds.

Electromagnetic Coupling

The antenna on a tag serves two functions. It couples the tag electromagnetically to the reader enabling the tag to sense the interrogation by the reader, and harvest the power from the electromagnetic radiation emitted by the reader. Second, it enables a tag to transmit data to the interrogating reader.

One of the simplest antennas is a coil that is *inductively coupled* to a similar coil in the reader. The time-varying magnetic flux generated by the reader's coil induces a current in the tag's coil, which can be rectified to power the chip's circuitry. Such transformer-like coupling is effective if the reader and tag are sufficiently close—that is, the tag is in the *near field* of the reader's antenna. The general rule of thumb is that the tag is considered to be in the near field of the reader if the distance between the reader and tag is less than the wavelength of the radio wave used [Banks et al. 2007]. For instance, a 13.56 MHz radio wave has a wavelength of about 22 meters. A tag is considered to be in the near field of a reader transmitting at 13.56 MHz, if the reader-tag separation is less than about 22 meters. For 915 MHz communcation (UHF) the near field range is about 32 centimeters.

UHF tags with large reader-tag separation rely on *backscatter coupling*. In backscatter coupling the reflection cross section of the tag's antenna is modulated by the data to be transmitted by the tag leading to data-dependent variations in the power of the reflected wave. When the electromagnetic radiation reflected from the tag reaches the reader the power variation in the reflected wave is demodulated by the reader to extract the data being transmitted by the tag.

If the chip relies completely on the radio wave energy harvested by the tag's antenna, then the range for reader-tag communication is also determined by the transmission power of the reader, the power requirements of the chip and the gains of the reader and tag antennas. For a more detailed discussion of the reader and tag antennas, see Finkenzeller [2010].

Types of RFID Transponders

RFID tags are classified broadly as *passive tags, semipassive tags* and *active tags* [Garfinkel and Holtzman 2006]. In a passive tag the energy needed to run the chip and to power the return radio transmission are derived from the reader's radio waves. A passive tag has no energy source of its own. At the other end of the spectrum, an active tag is powered by a battery housed on the tag. An active tag can broadcast its signal without having to be woken up by an interrogating reader, which means that both its chip and its radio transmission are powered by the on-board battery. Intermediate between passive and active tags is a semipassive tag which has an on-board battery to power the circuit. A semi-passive tag does not broadcast its signal until it is woken up by the electromagnetic waves from the reader. The energy for the tag's radio transmission back to the reader is still derived from the radio waves coming from the reader [Banks et al. 2007].

EPCglobal, an organization that is working to promote global standards in RFID, has suggested that tags be classified into six classes, Class 0 to Class 5 [Banks et al. 2007]. Tags of Classes 0, 1, and 2 are the basic passive backscatter tags with short range. Class 0 tags have read-only memory, Class 1 tags can be written onto once and the memory of Class 2 tags can be rewritten multiple times. Tags of Classes 3, 4, and 5 on the other hand are battery-assisted and able to support sensors. While Class 3 tags are still backscatter tags, relying on the energy from a reader's radio waves for data transmission, tags from Classes 4 and 5 are capable of active transmission.

TABLE 2.1

Frequency Bands for RFID Tags

RFID Frequency Class	Center Frequency
Low frequency (LF)	125 KHz
High frequency (HF)	13.56 MHz
Ultra high frequency (UHF)	915 MHz
Microwave frequency	2.45 GHz

The tags can also be classified by whether the memory in their chips is *read-only* or *read–write*. In read-only tags the tag's serial number is burned into the tag by the manufacturer. On the other hand read–write tags allow users to enter object-specific information on the tag.

RFID tags are also designed to operate at different frequencies. Some of the common frequencies at which they operate in the United States are shown in Table 2.1. Generally, *low frequency* (LF) and *high frequency* (HF) tags are used when the reader-tag range is expected to be small. When larger reader-tag ranges are desired the *UHF* and *microwave frequency tags* are employed [Sweeney 2005].

Interference and Collision

The environment in which RFID readers and tags operate exerts considerable influence on the wireless communication link. For example, metal objects reflect radio waves while liquids absorb them. Therefore, the presence of metals and liquids between an RFID reader and tag can prevent sufficient RF energy from reaching the tag. Since many everyday objects such as laptops and juice bottles have large metallic and liquid content reflection and absorption of radio waves pose a significant challenge to reader–tag communication.

Besides the interference from metals and liquids, the RFID devices are also required to be tolerant of interferences from other devices with which they share the radio-frequency bands. The frequency allocations in the radio wave spectrum and the power limits on devices emitting in this spectrum within the United States are governed by the Federal Communications Commission (FCC) and can be found at [USDOC 2011].

Within the radio spectrum certain ranges, earmarked for *industrial scientific and medical* devices, form the ISM band [Sweeney 2005]. The frequencies in the ISM band are open for use by unlicensed devices [FCC 2002]. Devices operating in the ISM band are expected to be tolerant of interference from other devices operating in the band.

Forbidden from interfering with the radio-frequency broadcasts of licensed applications such as radio stations, TV stations and police communications, the RFID devices are constrained to operate in the unlicensed ISM band in the United States. The range over which the tag and reader can communicate depends on the frequency used in communication. For ranges under 1 meter, the LF and HF bands are used. For ranges over 1 meter, typically UHF and microwave frequencies are used [Finkenzeller 2010]. In the United States the UHF frequency band used for RFID communications is 902–928 MHz [Sweeney 2005].

RFID devices are not only expected to be tolerant of interfering communications from other devices using the ISM bands, but they are also required to perform what is called *frequency hopping* over the allowed band. That is, an unlicensed device broadcasting in an ISM band, such as the 915 ± 13 MHz UHF band, is required to randomly switch among the 124 frequency channels in the range 902–928 MHz, approximately every 200 milliseconds [Sweeney 2005]. The frequency hopping ensures that no channel within the band is hogged by a single device. Since a reader interrogating a tag could use any of the channels, tags are required to respond to the entire spectrum in the band.

When several tags and/or readers attempt simultaneous communications in a small region one encounters collision problems. Collision occurs when multiple tags, in the vicinity of a reader, respond to a reader's interrogation—leading to tag collision—or when multiple readers in the vicinity of a tag simultaneously interrogate a tag—leading to reader collision [Jiang and Yeh 2009]. Such collisions degrade reader–tag communications considerably, slowing down the infrastructure's performance.

Several anticollision algorithms have been developed to address the reader collision problems. If the readers in an environment are under central control, then the reader collision problems can be handled through Time Division Multiple Access. That is, the readers are assigned specific intervals in which they are allowed to interrogate. Readers that have the potential of colliding are assigned nonoverlapping time intervals. An alternative would be the Frequency Division Multiple Access in which the allowed band of frequencies is divided into different channels, with readers assigned separate frequency channels for reader–tag communication. A third strategy, termed Listen Before Talk, is the Carrier Sense Multiple Access, in which a reader checks if the frequency channel in which it wants to interrogate is free before starting its transmission.

Several anticollision algorithms have been developed to mitigate the tag collision problems as well. For example, in the ALOHA protocols each tag waits for a random interval of time after receiving the reader signal before it responds. These protocols do not ensure anticollision. The tree-based protocols iteratively partition the tags into nonoverlapping groups until each group contains just one tag. For example, a reader could broadcast a string that elicits response only from tags whose identifiers have prefixes that match the broadcast string. If multiple tags respond to such a request, then the reader could rebroadcast a longer string by appending an extra bit (0 or 1) to the previous prefix string. By iteratively growing the string and branching on it each time there is a collision, the reader could read all the tags in the vicinity. Similarly, the reader could employ counter-based protocols for avoiding tag collisions. Jiang and Yeh [2009] contains a detailed discussion of the above anticollision protocols.

Sensor Tags

RFID tags that house on-board sensors and can measure physical parameters in their environment are called *sensor tags.* In the EPCglobal classification, these tags would belong to Classes 3, 4, and 5. Currently available sensor tags are capable of measuring many environmental features. Notable among them are the sensor tags that can measure temperature, humidity, and light; detect chemicals and radioactivity; determine location (using GPS) [Banks et al. 2007]; and measure acceleration and inclination [Ruhanen et al. 2008].

Many of the sensor tags function as transducers that convert the changes in the environmental parameters into measurable electrical changes in the tag's circuitry [Ruhanen et al., 2008]. For example, the capacitive pressure sensors measure the change in the capacitance of a capacitor that is subjected to external pressure. The capacitance of a parallel plate capacitor being dependent on the separation between the plates is affected by the external pressure [Cho et al. 1990]. Similarly, change in intensity of ambient light can be measured through the strength of the photoelectric current induced by it. See Ruhanen et al. [2008] for details of operation and availability of various sensor tags.

Sensor tags play an important role in many applications. For example, temperature sensors help monitor the refrigeration of perishable produce. The pressure sensors have been used in automobile tires to detect anomalous drop in tire pressure. And acceleration sensors have found use in monitoring shocks to fragile items [Ruhanen et al. 2008].

Translation of RFID Data

RFID data has unique characteristics. The identifier stored on a tag, like the license plate identifier on an automobile, does not encode any information about the properties of the object to which it is attached. Thus, following the acquisition of the raw data from the tag—that is, the serial number stored on

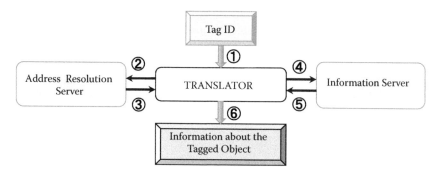

FIGURE 2.6
Translation of raw RFID data into information about the tagged object.

the tag—the data has to be *translated* into information about the associated object, as shown in Figure 2.6.

The details of the translation process depend on the RFID infrastructure. For example, the Ubiquitous ID network and the EPCglobal network—two of the prominent frameworks—do not use a common information architecture for the translation process. In the following discussion, we present the conceptual details of the translation process without referring to any particular RFID infrastructure.

Upon receiving the tag's identifier (①) the translator invokes the assistance of a service—called the Address Resolution Server (②). Given the tag's identifier, the Address Resolution Server returns the location—usually a Uniform Resource Locator (URL) on the World Wide Web*—at which the translator may find the information about the object associated with the given tag identifier (③). The Address Resolution Server could be implemented as a distributed service, patterned after the Internet's Domain Name Server.† Once the translator has the URL of the database—called the Information Server in Figure 2.6—at which the information about the tagged object resides, it can query the database (④) to obtain the information about the object (⑤). The information about the tagged object is then funneled to the application software (⑥).

Complex Events

In addition to the information about the associated object, often the time and location at which the tag is detected also contain useful information. The

---

* Uniform Resource Locator and World Wide Web are discussed in Chapter 4.
† Domain Name Server (DNS) is discussed in Chapter 3.

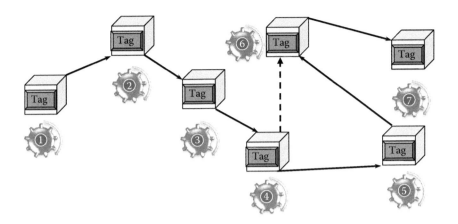

FIGURE 2.7
An example of a complex event in a multistep process.

following hypothetical example illustrates the utility of the time and location information.

Consider a seven-step process shown in Figure 2.7. The process could pertain to the analysis of a tagged biological sample in a laboratory or the assembly of a tagged electronic device by a human. The gears indicate the spatially distributed machines that must process the tagged sample/device in the order represented by the numbers.

The integrity of the process could be ensured by positioning RFID readers alongside each of the machines. As the tagged object is routed through the different steps the readers record both the serial number of the tag and the time at which the object was detected. The triple (reader ID, tag ID, time) represents a *simple event,* namely, the detection of a tag by a reader at a particular time. A time-ordered sequence of such simple events often encodes a *complex event.* A complex event is a pattern encoded in an ensemble of simple events, and by definition cannot be discerned by looking at the simple events separately and is detected only by a consideration of an entire ensemble of them. For example, the desired complex event in the above example is the correct routing of the tagged object through the seven-step process. The time-ordered sequence of triples encodes information about whether or not the complex event occurred. For example, if step 5 is inadvertently skipped, the missing data triple from Reader 5 signals that the seven-step process was not correctly executed, and that the desired complex event did not occur. Likewise, if the machines are visited in an order different from that intended, then the stream of data triples will signal the nonoccurrence of the complex event.

Processing RFID Data in Real Time

RFID infrastructures typically generate large volumes of noisy and incomplete data that often stream in at high velocity. The information processing system tied to the RFID infrastructure would be required to convert the incoming raw data into decision-aiding information often in real time. For example, an unusual movement of tagged objects in a warehouse, suggestive of possible theft in progress, would require the associated information infrastructure to detect the complex event and initiate a response in real time.

As is the case in the warehouse example mentioned above, the incoming data stream often has finite shelf life within which the information contained in it should be harvested. Such stringent real-time constraints make it impractical to store the incoming data streams in a database before processing them. Instead, the data streams need to be processed on the fly. Consequently, the traditional data processing paradigm based on *DataBase Management Systems* (DBMS), which insists on storing the data prior to processing it, is not viable for processing high-velocity RFID data streams. New data processing techniques are being developed based on the so-called *Data Stream Management Systems* (DSMS). Also, the traditional query languages, such as SQL, that are used to interact with databases, are replaced with languages, such as the *Event Processing Languages* (EPLs), that enable the detection and processing of events. The interested reader is referred to Eckert et al. [2011] for further details.

Smart Objects

Wireless tags serve to bridge the cyber and physical worlds by making tagged physical objects visible to the cyber infrastructure. As we discussed before, tagging an object is like attaching a license plate on a vehicle and does not endow the object with any *computational intelligence*. Tagging does not make an object *smart*. Merely tagging objects preserves the separation between the physical world comprising objects that lack computational intelligence and the cyber world in which all intelligent processing occurs.

In several applications, such as those discussed later in this chapter, the main interest is in making a large number of objects visible to the cyber infrastructure while minimizing the associated delay and costs. Simple wireless tags provide the optimal technology for such applications. The tag costs are kept low by storing the barest minimum amount of information on the tag. The downside of not putting intelligence in the physical objects is that the cyber infrastructure will have to contend with a deluge of unprocessed (raw) data. An alternative paradigm is to move some of the intelligence into physical objects to make them not only visible to the cyber infrastructure but also "smart." We briefly discuss such smart objects below.

There is no consensus on the definition of a "smart" object. As a working definition, we will take a smart object to be a *physical object that is endowed with computational intelligence and is visible to the cyber infrastructure*. This definition excludes intelligent devices that are not connected to the cyber infrastructure. The exclusion is justified by our focus on the connection between the cyber and physical worlds.

It is important to note the distinction between visibility and addressability. A tagged device is visible to the cyber infrastructure since its presence can be sensed by readers connected to the infrastructure. However, a tagged device may not have an identity in the cyber infrastructure and hence may not be addressable. In order for a device to be addressable it must be assigned a globally unique address in the cyber infrastructure, and must have the hardware connectivity and software capability to communicate with other addressable devices.

A prototypical example of a smart device is a smart energy meter [Vasseur and Dunkels 2010]. With the demand for electrical energy fluctuating over time, utility companies face the challenge of handling the peaks in the demand for electricity. Especially in a deregulated electricity market handling the peaks in demand could involve purchasing electricity at premium prices from fellow suppliers. The net profits of the suppliers can be improved by encouraging the consumers to shift their electricity usage from peak hours to off-peak hours to the extent possible. One strategy to encourage usage during off-peak hours is to tier the cost of electricity depending on the time of day with the cost of electricity being highest during peak hours.

The older electricity meters only record the total amount of electricity consumed during the billing period. Such coarse usage data is inadequate to implement tiered pricing. The smart energy meters provide the consumers as well as the electricity suppliers with the required fine-grained real-time data about energy usage. For example, a smart meter installed in a residence can be programmed to send an hourly update to the electricity supplier providing data about both the energy consumed and the time at which it was consumed. Such data is also helpful to those consumers who choose to shift their energy usage to off-peak hours to benefit from the tiered pricing.

The smart energy meters use one of several modes of communication to send periodic updates about electricity usage to the suppliers. Lately smart energy meters are IP-enabled and can use the existing Internet infrastructure for their communications. More detailed discussions on smart energy meters are presented in [Vasseur and Dunkels 2010].

Whereas smart energy meters improve the monitoring of energy usage, smart thermostats help minimize energy usage [Nest 2012]. Suboptimal settings of temperatures in homes lead to wastage of significant amounts of energy. It is estimated that a mere one-degree change in the temperature can translate to a 5% change in the energy consumed [Pogue 2011]. EPA estimates that a properly programmed thermostat can cut the energy used for heating and cooling by about 20% [Nest Labs 2012]. Currently available smart thermostats are equipped with the intelligence to sense whether a living space is occupied and adjust the temperature accordingly. Some of the smart thermostats can also be programmed over the Internet, giving owners the capability to monitor the conditions in the living spaces over the Internet.

Temperature is one of the many conditions that can be monitored over the Internet, thanks to an emerging suite of new technologies. The smart home security systems currently available on the market [Xfinity 2012] enables users to remotely control appliances—for example, turn lights on and off over the Internet—and get real-time streaming video of the living spaces.

These examples give a glimpse of the landscape of smart devices that are appearing on the market. Most of the current smart devices are focused on existing needs and in that sense belong to the first wave of applications of smart technologies. The second wave of applications, the new unanticipated possibilities, will likely emerge with the development of the capability to build large networks of smart devices that can interoperate without human intervention.

Smart Networks

While individual smart devices are of interest in their own right, networks of smart devices are of greater interest as the synergistic interactions among devices could give rise to nontrivial collective behavior. An example of such smart networks is the proposed *Vehicular Ad Hoc NETwork* or VANET for short. The objective of a VANET is to use the moving vehicles on the roads, say in a city, to forge a dynamic mobile network. The vehicles serve as nodes of the network and are expected to have the capability to communicate with other nearby nodes in the network. Communication between distant nodes—nodes that are too far apart to communicate directly—occurs in a multihop mode. That is, the messages between communicating end nodes are relayed by intermediary nodes in the network.

An example of a nontrivial behavior that can emerge from the synergistic interactions of vehicles is the possible self-organization of traffic. Assuming that all vehicles are equipped with built-in GPS, a VANET enables each vehicle to periodically broadcast its location to other vehicles in the network. Such network-wide real-time data provides every vehicle adequate information to determine where the traffic is congested. The navigation systems of the vehicles can then dynamically reroute the vehicles to avoid congested traffic [Padron 2009].

The monitoring of energy usage by a utility company and the dynamic self-organization of traffic represent two contrasting modes of control called, respectively, the *orchestrated control* and *choreographic control*. In the orchestrated control mode a group of devices operate under the direction of a central controller, just as a group of individual performers in an orchestra play under the direction of a conductor. The smart energy meters, operating under the control of their utility company, provide an example of orchestrated control. In contrast, the choreographic control mode is characterized by the lack of central control. The term itself is suggestive of the manner in which the choreographed movements of an individual dancer in a performance are governed, not by a conductor, but rather by the movements of the other dancers. Likewise, in a choreographic control mode the behavior of a smart device is governed by that of the other smart devices with which it interacts. The VANET discussed above is an example of a system operating with choreographic control. In congestion avoidance, the movement of a vehicle is determined by the movements of the other vehicles in the network.

The above example of self-organization of traffic illustrates how the interactions among smart objects, especially when they are operating in choreographic control mode, can give rise to nontrivial emergent behavior. *Emergent behavior*, discussed in Chapter 12, is a fascinating phenomenon that arises pervasively in systems involving smart components.

Enabling heterogeneous smart devices to interoperate with each other would be a critical first step in building intelligent adaptive systems. The *IPSO Alliance*, discussed further in Chapter 6, is a multiorganization collaboration dedicated to promoting interoperability among smart devices.

Applications

In the following sections we present some applications of the RFID technology. The commercial interest in RFID is still largely focused on its application to supply chains. Our interest in RFID technology is broader in scope and envisions RFID as a bridge technology for I-2. Accordingly, we have selected examples that showcase innovative uses of RFID outside the context of supply chains.

The RFID technology provides two key capabilities: (1) enhanced visibility, that is the capability to detect tagged objects that are not necessarily in the line of sight, and often not even in the vicinity, and (2) rapid detection, that is, the capability to detect objects at high speeds, as the technology does not require human intervention. The applications we discuss below exploit one or both of the above capabilities.

Enhanced Visibility of Objects

The following applications exploit the enhanced visibility that is made possible by the RFID technology.

Tracking Lost Children in Amusement Parks: Locating lost children is a problem that parents routinely face in large amusement parks. A study by Intimetrix, over a 12-month period, showed that about 27% of the parents visiting large amusement parks lost one or more children for at least brief periods during their visits [Dver 2007]. Large malls and stores report similar alarming statistics about lost children [SentryGPSid 2009]. RFID-based wristbands have been successfully used to provide real-time information about the locations of children in large amusement parks [Sullivan 2004]. For example, visitors at the 140,000-square-foot theme park in Fort Lauderdale are issued RFID-enabled wristbands upon entry. Readers located throughout the park track the signals from these wristbands every few seconds and funnel the information to a central computing system, which maintains a constant awareness of the locations of all the wristbands. Parents can obtain the instantaneous locations of their children by interrogating the central computing system using terminals distributed throughout the park.

Mining: Blasting sessions in mines have to be coordinated with worker movements within the mines to ensure that all miners are evacuated from a region before blasting begins in it. In addition, timely evacuation during emergencies also hinges on maintaining an awareness of the locations of the miners. Tracking workers' movements had posed challenges previously, exposing miners to avoidable risks. RFID technology is being successfully used to address these problems in the Paardekraal mine in South Africa. The RFID tags are bundled with the lamps that miners use underground. Tracking the tagged lamps using a local area network of readers has enabled the infrastructure to track the miners and equipment effectively [Violino 2005].

Baggage Handling: Baggage handling in airports involves transferring the passenger bags to either connecting flights or to baggage claim terminals for pickup. Until 2004, the Hong Kong International Airport, which handled about 40,000 bags daily, relied on bar code based tags for handling baggage. The infrastructure was upgraded at a cost of about $6.5 million to RFID-based routing. Now RFID readers located along the network of conveyor belts interrogate the tagged bags and make routing decisions based on the destination information associated with the tags. The airport reports that the

RFID-based routing has enabled it to increase its baggage handling capacity by about 5%, while the accuracy of the system has increased from about 80% for bar code-based tags, to about 97% for RFID-based tags [Swedberg 2009].

Document Tracking: Although documents are increasingly being stored in digital formats, physical files continue to be used extensively in modern offices. In settings that involve heavy traffic of physical files, the handling of files is fraught with two problems. Locating a misplaced physical file often requires considerable manual effort. Second, obsolete files are not expunged regularly. As a result, the office space is used inefficiently. Using RFID technology for file handling in education, insurance, legal, medical, and military and government settings has led to gains in efficiency in the use of both human and space resources [3M 2012, RFID Update 2007].

Containers: The maritime ports in the United States receive more than 11 million containers every year [CSI 2011]. The container traffic is even heavier at busier ports like the Port of Singapore, which is estimated to handle about a seventh of the global container traffic [PSA 2012]. Only an estimated 5% of all the containers arriving at the United States' ports are physically inspected [Banks et al. 2007], making the containers a potential source of threat to national security [Flynn and Kirkpatrik 2006].

Previously, conventional seals were being used to secure containers. The problem with using conventional seals is that checking their integrity—to see if they have been tampered during transit—requires manual inspection of the seals. The large volume of container traffic makes it infeasible to manually inspect all of the containers flowing through a port.

Recently, smart RFID seals are being used as a high-throughput alternative to the conventional seals. The smart RFID seals make it practical to quickly monitor the integrity of all of the containers, not only at the ports but also at intermediate points along the transit routes. Several ports across the world, including the Port of Singapore, are now using RFID seals on containers bound for the United States as part of the Container Security Initiative. In addition to RFID seals, smart sensor tags are also being used to detect changes in conditions, such as temperature, pressure, and humidity, inside a container and detect the presence of radioactive, biological, and chemical agents in containers. Bundled with GPS, RFID seals and sensor tags are providing unprecedented tracking capability, which is being used by the United States Department of Defense for high-security containers [Banks et al. 2007].

Tire Pressure: Deflated tires not only adversely impact the fuel efficiency in vehicles but also pose risks to the safety of the passengers. While extreme deflation or inflation of tires is more readily detected by visual inspection it is often harder to detect smaller deviations from recommended tire pressures. Further, the regimen of periodic manual measurement of tire pressures is unreliable and fraught with large variations among the vehicle operators. The RFID technology is being successfully deployed to meet the objective of raising both fuel efficiency as well as safety through continuous monitoring of tire pressure. For example, Michelin is supplying RFID-enabled tires to some of Stagecoach London's buses [Swedberg 2012]. The RFID tags are coupled to wireless pressure sensors which measure the air pressure in tires. The pressure information is then transmitted wirelessly providing accurate real-time information about tire pressures.

Theft protection: RFID technology is also being used within automobiles to prevent theft. It is estimated that nearly half of all the vehicles manufactured in the United States are equipped with anti-theft devices that rely on RFID technology. The annual sales of antitheft devices in the United States in 2005 were estimated to be about $4 billion [Banks et al. 2007].

Rapid Detection of Objects

The RFID technology enables a rapid detection of tagged objects, which is exploited in many applications of the technology. The following paragraphs present a few illustrative examples of such applications.

Electronic tolls: Electronic collection of tolls is one of the prominent applications of RFID. Toll booths, especially on roads that support heavy traffic, are bottlenecks that significantly slow down the traffic. Congested traffic has an associated cost tied to the time and fuel wasted in slow traffic. In addition, staffing the toll booths carries personnel costs. RFID-based electronic toll collection provides a remedial alternative that cuts costs both for commuters and toll collectors. In RFID-based electronic toll collection, RFID readers installed in toll booths wirelessly retrieve the serial numbers of the RFID transponders in the passing vehicles. The serial number of a transponder is then used to access the owner's account for electronic collection of toll. Since the toll collection is delegated to the back-end computing facility, and the only transaction that has to occur on the road is the electronic detection of the transponder's serial number, the vehicle is not required to stop or even slow down for toll collection. The electronic toll collection is one of the pervasively used applications of RFID that is impacting the lives of millions of commuters annually [Banks et al. 2007].

Retail: It is estimated that the retail stores lose about $30 billion annually in missed sales because of products not being on the shelves when the customers look for them [IBM 2004]. In addition, the estimated global losses due to inventory shrinkage—resulting from a combination of factors such as human errors, inaccurate inventory, and shoplifting—is about $119 billion. Further, the shrinkage rate increased by about 6% year-to-year in 2011 [Roberti 2011b, Banks et al. 2007]. Item-level tagging of products has been found to reduce losses due to both missed sales and inventory shrinkage.

Item-level tagging helps stores electronically interrogate the shelves and thus gather critical real-time information about the inventory and trends in sales. Such real-time monitoring enables stores to lower the incidence of missed sales. Electronic tagging also reduces errors due to incorrect shelving or misplaced items, as displaced items are more efficiently detected by readers placed near shelves than by periodic visual inspection by humans. Recognizing the value added by item-level tagging, a German clothing manufacturer has tagged more than 26 million articles of clothing with RFID tags since January 2011 [Gerry Weber 2012]. Item-level tagging adds overheads to the costs and raises privacy concerns. However, the economic magnitude of shrinkage and the rising rate of shrinkage are providing compelling arguments for widespread adoption of RFID technology in the retail sector.

Temperature Sensing: The shipment of temperature-sensitive biological and pharmaceutical products to distant destinations presents a unique challenge. Bio-molecules like proteins exhibit their desired functionality only inside certain molecule-specific ranges of temperatures. If exposed to temperatures outside the recommended ranges they are denatured, that is, lose their functionality. Therefore, biological products such as vaccines that contain temperature-sensitive molecules need to be maintained within recommended temperature ranges during their transportation. Temperature-regulated containers are used for transporting such products. However, if the ambient temperature inside such containers vary outside the tolerable range, either due to equipment malfunction or human error, then customers need to be alerted in real time to enable an efficient response to the adverse event. RFID-based temperature sensors are being used by carriers like DHL to provide customers the ability to track the shipment conditions almost in real time [DHL 2007]. Sensors installed inside the containers provide information about the temperature within the containers to readers located at selected points along the transit. The readers in turn route the information to the customers. With the growth of the biotechnology and pharmaceutical industries, the volume of shipment of temperature-sensitive products is expected to soar in the coming years. RFID-based innovative solutions, such as those being employed by DHL, are providing the service needed for the emerging needs.

Contactless Transactions: Magnetic stripe cards, such as credit cards and identification cards, are not very versatile. The functionalities of multiple cards are being bundled together into a single contactless smart card using RFID technology. For example, the Octopus card that has gained popularity in Hong Kong doubles as a smart card for electronic transactions and also for controlling access into buildings [Octopus 2012].

Challenges in Healthcare Delivery

The healthcare sector in the United States is confronted with two main challenges: (1) *containing the escalating healthcare costs* and (2) *improving the quality of healthcare delivered to patients.* The reports from the Institute of Medicine titled "The Healthcare Imperative: Lowering Costs and Improving Outcomes" [IOM 2010], and "To Err Is Human: Building a Safer Health System" [IOM 2000], and the references cited therein showcase the challenges.

Escalating Healthcare Costs: The annual healthcare expenditures in the United States in 2009 were estimated to be about $2.5 trillion, or about 17% of the gross domestic product of the United States [IOM 2010]. The healthcare expenditures are projected to reach about $4.4 trillion by 2018, or about 20% of the national GDP [CMS 2009]. The rising healthcare expenditures do not appear to be affected by economic downturns. For example, Bureau of Labor Statistics [BLS 2009] reports that between August 2008 and August 2009, the consumer price index decreased about 1.5%, even as healthcare costs increased by about 3.3% over the period.

As staggering as these rising costs are, an even more startling statistic is that about 30% of the healthcare expenditures in 2009, or about $765 billion out of the total expenditure of $2.5 trillion, appears to be *wasteful expenditures* [IOM 2010a]. That is about 5% of the national GDP is being spent on wasteful expenditures in the healthcare sector. The statistics on wasteful expenditures reported in [IOM 2010a] are summarized in Table 2.2.

TABLE 2.2

Wasteful Expenditures in Healthcare Delivery

Category	Contributors	Wasteful Expenditure
Unnecessary services	• Excessive use of services • Defensive medicine • Needless high-cost services	$210 billion
Excessive administrative costs	• Duplicative costs for insurance • Unproductive documentation	$190 billion
Inefficiently delivered services	• Medical errors • Uncoordinated care • Inefficient operations	$130 billion
Excessive prices	• Uncompetitive product prices • Excessive variation in service prices	$105 billion
Fraud	• In Medicare and Medicaid claims	$75 billion
Missed prevention opportunities	• Poor delivery of prevention	$55 billion
Total wasteful expenditure		**$765 billion**

Some of the wasteful expenditures, such as excessive use of services, are rooted in decision making by patients and doctors. However, aspects such as inefficient operations and unproductive documentation point to systemic inefficiencies in the healthcare delivery infrastructure.

The report also estimates that over the next 10 years, about $181 billion can be saved by streamlining administrative costs, about $80 billion by improving hospital efficiency, about $12 billion by preventing medical errors, about $10 billion by preventing fraud, and about $9 billion by shared decision making. That is, about $292 billion can be saved over the next 10 years by improving the process of healthcare delivery.

Return on Investment: Although the United States spends the largest amount per capita on healthcare among all industrialized nations [Peterson and Burton 2008] in measurable outcomes such as life expectancy and infant mortality the country appears to be worse off than a few other nations [Anderson and Frogner 2008; Docteur and Berenson 2009]. The IOM report [IOM 2010] indicates that workers and employers in the United States spend about 58% more on health care than those in other industrialized nations and yet the general health of the U.S. workforce is worse off by about 10%. Emerging economies like Brazil, India, and China [IOM 2010, Milstein 2009], on the other hand, spend only about 15% of the amount the United States spends on health care and yet the health of their workers appears to be only about 5% worse than that in the United States.

Clearly, many hidden variables influence the health measures such as life expectancy. Whatever the contributing factors may be the obvious inference that these statistics seem to be suggesting is that the United States is spending more, per capita, on health care than other industrialized nations, and yet the workforce in the United States appears to have poorer health than in the other developed countries.

Quality of Healthcare: The IOM report "To Err Is Human" [IOM 2000] estimates that every year at least 44,000 deaths occurring in the United States are due to medical errors (the estimate is based on an extrapolation of the findings in studies done in Colorado and Utah). The report puts the estimate in context using CDC statistics, which show that the number of deaths in the United States in 1997 due to motor vehicle accidents was 43,458, due to breast cancer was 42,297, and due to AIDS was 16,516 [CDC 1999]. Medication errors alone are estimated to be responsible for about 7000 deaths annually in the United States. In contrast, about 6000 deaths are attributed to injuries in the workplace every year [Phillips et al. 1998, IOM 2000]. The studies by Bates et al. [Bates et al. 1997, IOM 2000] showed that the average annual costs due to preventable adverse drug events in a 700-bed hospital were about $2.8 million. Extrapolation of the estimate suggests that about $2 billion are being wasted every year due to preventable adverse drug events.

RFID technology is being increasingly used to improve the efficiency of healthcare delivery while reducing costs and medical errors. In the following paragraphs we present selected examples that illustrate the applications of RFID technology in the healthcare sector.

RFID-Enabled Healthcare Delivery

Mobile medical equipment: Stretchers, wheelchairs, portable intravenous poles, defibrillators, oxygen tanks, infusion pumps, and monitors are examples of mobile medical equipment that are routinely used in healthcare facilities. The sharing of mobile medical equipment by several interacting units within a healthcare facility poses unique challenges. The healthcare professionals often need the mobile equipment at a short notice. Second, the equipment must be maintained in a sterilized state of readiness to permit its use at a short notice. Since they are mobile, and used for varying periods of time, it is difficult to centrally track the location and sterilization status of different pieces of equipment. The lack of such real-time information often encourages healthcare professionals to hoard pieces of equipment that are critical for their operations [Roberti 2006], further diminishing the duty cycle of the shared equipment. Aside from the inefficiencies tied to the wasted man hours spent searching for the equipment, the delays in locating the equipment could impact the quality of healthcare delivered to patients. RFID technology has been successfully employed to mitigate the problems associated with shared use of mobile medical equipment. Tagging the pieces of equipment enables a network of readers located throughout a hospital to track the location of the equipment. The readers feed the data into a central computer network which maintains the real-time information about the location and sterilization status of each piece of equipment. A healthcare professional seeking a piece of equipment can determine its location from any terminal that is connected to the facility's Intranet. Such *Real Time Locating System* (RTLS) for mobile medical equipment is being used in Memorial Hospital, Miramar, Florida [Violino 2010].

Retained Surgical Items: A related serious problem pertains to the so-called *retained surgical items,* or surgical instruments and consumables that are inadvertently left behind inside a patient during surgery. The protocol used to guard against the adverse event of leaving an instrument or a consumable inside a patient is to perform a manual count of all the pieces of equipment and consumables before completing the surgery. Such a manual count, apart from being error-prone, is time-consuming and expensive given that the cost to run an operating room is of the order of hundreds of dollars per minute. Adverse events related to retained surgical items imperil the patient safety as well, necessitating avoidable exposure to x-rays and possibly even remedial surgery to retrieve the items. RFID technology provides an elegant and reliable solution to the problem. Surgical instruments as well as consumables such as sponges are tagged enabling the readers located in the operating room to track the tagged items. Using readers of different ranges and appropriate control software the problem of retained surgical items is greatly mitigated by the RFID technology [Swedberg 2010]. For example, in a survey of 2,961 cases in which surgical sponges were tagged with RFID, it was found that in 21 cases the sponges were inadvertently left inside the patients during surgery [Roberti 2012].

Newborn Infants: More than half of the 233 children abducted in the United States between 1983 and 2004 were taken from healthcare facilities [Collins 2005]. Ensuring the safety of newborn infants during postpartum care is a problem that healthcare facilities face. RFID technology is being successfully used in facilities such as Lucile Packard Children's Hospital in Palo Alto and Doctors Hospital in Dallas, Texas [Orlovsky 2005]. At birth, an infant is tagged with an anklet containing an active RFID transponder and the mother with a bracelet tag that has a matching serial number. The infant's tag is programmed to send a signal to the hospital's readers, periodically enabling the network of readers to alert the staff if the tag malfunctions or if someone tries to tamper with the tag. The elevators and exits of Lucile Packard Children's Hospital are equipped with readers which raise an alarm if they detect an infant's tag without the matching tag of the mother.

Hospital Beds: Hospital beds are precious resources in healthcare facilities that offer in-patient care. For example, St. Vincent's group, which manages about 67 acute-care hospitals in 20 states provides in-patient care to about 17,000 patients every year [Gambon 2006]. When hospital beds are unavailable patients have to be diverted to other hospitals. In 2004, it was estimated that St. Vincent's lost about $20 million in possible revenue on account of diverting patients to other hospitals.

St. Vincent's estimates that it used to take about 6 hours to enter a patient's discharge information into the computing system. Consequently, for about 6 hours after a patient's discharge the system would not show the availability of a vacated bed even as patients waited for beds and were possibly diverted to other hospitals for apparent nonavailability of beds. The RFID technology has cut down the lag in updating the hospital's records from about 6 hours to under 6 minutes at St. Vincent's. Patients charts were tagged with RFID transponders, helping the hospital track the movement of the patients (as charts accompany patients) and also expedite the entry of the patient's discharge information into the hospital's information database. The increased visibility of the status of hospital beds translates to improved revenue and better patient care.

Pharmaceuticals: Pharmaceutical drugs are small high-value products, which are often shipped across national boundaries. The global distribution network often involves as many as ten handoffs of a product as it makes its way from the manufacturer to the eventual consumer. Consequently, the distribution network is vulnerable to interception by counterfeiters who could replace the original product with counterfeit drugs. The counterfeit drugs are made to resemble the original drugs in appearance making it difficult to detect the interception. It is estimated that the pharmaceutical industry loses about $40 billion worldwide annually in lost sales on account of counterfeit drugs. The World Health Organization estimates that nearly 6% of all the drugs circulating globally are counterfeit [Paddison 2004]. In addition, the counterfeit drugs drive up the insurance costs as they may not be as effective as the original products [Roberti 2011a]. RFID tags are now being used to fight interception of the pharmaceutical supply chain by counterfeiters [IBM 2007]. The bottles containing the drugs are being tagged with RFID transponders. Readers placed at strategic checkpoints in the pharmaceutical supply chain then make it difficult for counterfeiters to intercept the supply chain [Paddison 2004].

Economics and Trends

The RFID market was estimated to be about $5.35 billion in 2010, representing a 15% increase from 2009. The application that had the biggest share of the RFID market was automobile theft protection discussed above. Emerging applications such as animal ID and baggage handling are expected to outpace the growth of older applications such as automobile immobilization and electronic toll collection in the years ahead. The older applications currently constitute about 61% of the RFID market and are expected to show a growth rate of about 6% annually. In contrast, the newer applications are expected to grow at a rate of about 19%. By 2014, the market is predicted to exceed $8.25 billion, with applications such as real time location showing the most rapid growth [ABI 2010].

Summary

The RFID technology enables us to digitally enhance the physical objects to make them visible to the cyber infrastructure. Variants of the basic RFID tags, such as the sensor tags, enable the cyber infrastructure to wirelessly acquire data about several physical parameters pertaining to the objects and the environment. The RFID and sensor technologies are currently being used to bridge the cyber and physical worlds in isolated settings and are expected to play a prominent role as bridge technologies in the emerging I-2 infrastructure.

   The discussion in this chapter presented a coarse-grained introduction to the hardware aspects of the RFID technology, and several examples that demonstrate how the technology is currently being applied to increase the visibility of the physical objects to the cyber infrastructures. Besides RFID, several other competing bridge technologies, such as optical sensors, are also expected to participate in the I-2 infrastructure. The details of the specific bridge technologies employed in I-2 is of limited relevance to the architecture of I-2. We restrict our discussion of bridge technologies to the foregoing discussion of the prominent representative of bridge technologies—the RFID technology.

Section III

Cyber Infrastructures

Overview

The bedrock for the current cyber infrastructure is the global data transport network called "the Internet." The Internet comprises nodes (devices with IP addresses), the communication links that connect the nodes and the protocols that govern the behavior of the network. One of the most important services running on the Internet is the World Wide Web, hereafter called just "the web." The web can be viewed as a large distributed library of interlinked digital resources residing on Internet's nodes, together with a user-friendly interface that facilitates navigation through the network of resources. The service provided by the web has contributed significantly to the widespread use of the Internet. In Chapters 3 and 4 we present an overview of the Internet and the web.

Earlier, the Internet comprised hardware nodes that were stationary. With the advent of mobile devices such as mobile phones and handheld digital assistants it became possible to connect mobile devices to the Internet, giving birth to an enhancement of the Internet called the "mobile Internet." Mobile Internet is of interest in the context of the I-2 since it embodies some of the essential features of I-2—namely, the wireless bridge between a mobile device and the cyber infrastructure. In Chapter 5 we discuss the mobile Internet and mobile web.

As we mentioned in the Preface, the infrastructure that is envisioned to integrate the cyber and physical worlds is called "the Internet of Things" in the literature. In Chapter 6 we present an overview of the work that has been done to date toward building the Internet of Things. The discussion in Chapter 6 provides the backdrop for our discussion of the I-2 infrastructure in Chapters 10 and 11.

3

Internet

The bedrock of the current cyber infrastructure is the Internet. In the following sections we present a brief overview of the Internet's architecture and history, focusing in particular on those aspects that are relevant to the later discussion of the I-2 infrastructure. A reader seeking a more complete discussion of the Internet is referred to [Comer 2006, Gralla 2006, Leiner et al. 1997]. Both the architecture of the Internet as well as the strategic decisions that its architects made to facilitate its growth embody important lessons about building successful distributed global infrastructures. The discussion in this chapter is a preamble to the discussion in Chapters 7 and 8 in which we discuss the guidelines for architecting the emerging distributed global infrastructure—the I-2.

A Simple Analogy

The Internet is a globally distributed infrastructure that provides electronic real-time data transport service. It bears some resemblance to the globally distributed postal service infrastructure. Whereas the postal service transports packages, the Internet transports digital packets called *IP datagrams*. Just as an entity has to have a postal address to be able to receive a postal shipment, an end node in the Internet has to have an *IP address* in order to be able to receive an IP datagram. The global postal service network is not controlled or operated by a single agency. Instead, it comprises a network of independently operated postal service infrastructures—one in each country—that cooperate to provide seamless global service. The Internet too is not owned or managed by a single entity. It is a global infrastructure that relies on the cooperation of several independently managed networks. A postal shipment typically goes through a multihop journey visiting several intermediate processing and distribution centers before reaching the destination. An IP datagram is also typically routed through several intermediate routers in its journey.

Although the Internet has several idiosyncrasies that distinguish it from the simple postal service—and we discuss the details of such idiosyncratic features below—the simple analogy, described above, might serve as a useful backdrop for the discussion in the remainder of this chapter.

Architecture of the Internet

The architecture of the Internet is best described by tracking a hypothetical online transaction. Accordingly, we consider a user who seeks to access a remote web page from a computer at his house. We will assume that he has multiple computers that connect to the Internet through a single gateway in his house. A *gateway* is a connection point between the user's home network and the external Internet. All communications between the Internet and the user's home network flow through the gateway. The user starts by specifying a web address—say http://www.purdue.edu/index.html—to his *browser* instructing it to retrieve the web page at the specified address. His request triggers a sequence of communications, which are shown ordered alphabetically in Figure 3.1. The events triggered by his request are outlined below and elaborated further in the sections that follow.

1. When the user's computer connects to the home network the gateway assigns an internal IP address to the user's computer, using *DHCP—Dynamic Host Configuration Protocol.* (For simplicity, we have assumed that the gateway also serves as a DHCP Server for his home network.)

2. The web address *http:www.purdue.edu/index.html* is forwarded by the browser, through the gateway, to a *DNS Server* at the local *Internet Service Provider* (ISP).

3. After a series of delegated requests, the local ISP's DNS Server returns the IP address of the *web server* corresponding to the address www.purdue.edu.

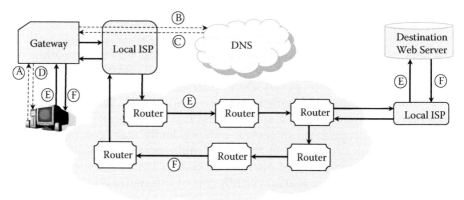

FIGURE 3.1
Coarse-grained illustration of the Internet architecture.

4. The browser then sends a request for the web page through the local ISP to the destination web server. The request is sent using the *TCP/IP protocol.*

5. The request is sent as a stream of one or more data packets. Each of the packets follows its own path through a global network of *routers* toward the destination web server.

6. When the destination web server receives all the data packets, it processes the request, retrieves the requested web page—a *HTML file*—packages the file into data packets, and sends each packet toward the IP address of the gateway, again using the TCP/IP protocol.

7. The packets flow through the user's local ISP and the gateway in his house to his computer guided by the gateway's *Network Access Table.*

8. The browser on the user's computer processes the HTML file it receives and displays the contents of the file.

The hardware resources, services and protocols that are mentioned above are discussed in greater detail below.

IP Address

An *IP* (Internet Protocol) *address* is a globally unique address assigned to a device to make the device addressable on the Internet. It is analogous to the street address of a house. Just as postal mail can be delivered to a house that has a street address digital data can be delivered on the Internet to a device that has an IP address. A device that does not have an associated IP address does not have an identity on the Internet.

Initially, devices on the Internet were assigned 32-bit IPv4 addresses. The IPv4 scheme permits about 4.3 billion devices to have addresses on the Internet. The emerging I-2 requires a much larger address space. Currently, the IPv4 is being replaced by the 128-bit IPv6 system, which provides enough address space for the foreseeable future.

Although the IPv4 is being phased out, it is instructive to understand the IPv4 addressing scheme. The 32-bit address in IPv4 can be logically partitioned into the *network address* and the *host address*. The host address specifies the address of the particular device within the network specified by the network address. In the recent *Classless Inter-Domain Routing* (CIDR) scheme [Fuller et al. 1993], the number of bits earmarked for the network address is specified by a number following a slash. Thus, for example, the IPv4 address

192.168.2.15/16 is interpreted as follows.* The first four numbers, which are separated by dots, represent the values of the four bytes in the 32-bit address. Each of the numbers, represented using 8 bits, lies in the range 0–255. The number/16 indicates that 16 most significant bits—namely, the bits corresponding to 192.168—specify the network address. The remaining two bytes—namely, 2.15—specify the address of the host device within the network numbered 192.168. The CIDR scheme enables a more efficient utilization of the available address space in the IPv4 scheme.

Static and Dynamic IP Addresses

Not all of the devices that connect to the Internet are assigned permanent or *static* IP addresses. For example, several computers in a user's house could connect to the Internet through a single gateway. The gateway has a globally unique IP address tied to it, and it dynamically assigns private IP addresses to the devices connected to it, using *Dynamic Host Configuration Protocol (DHCP)*. See [Droms 1997]. The advantage of assigning dynamic IP address to a device is that when the device is inactive the IP address can be assigned to a different device, enabling a large number of devices to connect to the Internet through the gateway. The private IP addresses assigned using DHCP do not have an identity on the Internet. For example, an IP address of the form 192.168.x.y, where x and y are in the range 0–255 represents an address on a private network, and not an address on the Internet.

Network Address Translation (NAT) Table

The communication between a computer with a private IP address and the external Internet is managed by address translation within the gateway as described below [Srisuresh and Holdrege 1999]. Several processes running on the user's computer could make simultaneous requests to access the Internet. When a process on the user's computer wants to communicate with some external node on the Internet, say a web server, the user's computer, which we will assume has been assigned the internal IP address 192.168.1.100 by the gateway, creates a port for the process, say port numbered 3000. A *port* is a virtual dock through which a process can receive

---

* The address 192.168.0.1 used in this example is not a real address on the Internet as the address 192.168.x.y is earmarked for private IP addresses in local area networks.

data. The computer then transmits the communication, intended for the web server, to the home gateway along with the internal *socket address*—that is, the pair (192.168.1.100, 3000). The gateway, which we will assume has IP address P.Q.R.S on the Internet, assigns its own port to this communication, say port numbered 3300. The gateway records in its *Network Address Translation (NAT) Table* that the external socket address (P.Q.R.S, 3300) corresponds to the internal socket address (192.168.1.100, 3000). The gateway then transmits the communication, together with the socket address (P.Q.R.S, 3300), to the destination web server, over the Internet. If the web server sends any communication to socket address (P.Q.R.S, 3300) then the NAT Table is used to direct the incoming external communication to port 3000 of the user's computer, and thence to the process that initiated the communication. For a more detailed description of the NAT-based mapping, the reader is referred to the discussion of IP masquerading in Chapter 5.

Domain Name Service

The Domain Name Service (DNS) translates a human-friendly address of a server, such as www.lib.purdue.edu, to the numeric IP address of the server, namely, 128.210.126.182. A website name such as www.lib.purdue.edu is like the name of a person, while its IP address is like the person's telephone number. The DNS acts as a global directory service that, given a web address, returns the associated IP address. See [Dostalek and Kabelova 2006] for a detailed discussion of DNS.

A server's address such as www.lib.purdue.edu is a hierarchical sequence of labels separated by dots. Each label is allowed to be 63 characters long. The labels are organized in the DNS database in a hierarchical tree structure as shown in Figure 3.2. Each node in the tree corresponds to a label. The node at the top of the tree is denoted with a dot and is called the *root*. Each subtree in the DNS hierarchy is called a *domain*. Thus, the entire DNS tree is the *root* domain. The subtree, including the *edu* node and all of its descendants, is the *edu* domain. The *purdue.edu* domain is shown boxed in the figure.

At every node in the DNS hierarchy is a *name server* that stores a table of mappings from certain host names to their numeric IP addresses as well as pointers to the name servers at its daughter nodes. The daughter nodes of the root are called the *Top Level Domains* or *TLDs*. The name server at the root stores the addresses of the name servers at the TLD nodes.

If a computer seeks to retrieve the numeric IP address corresponding to www.lib.purdue.edu from some remote location in the world, say India, then the following cascade of events unfolds.

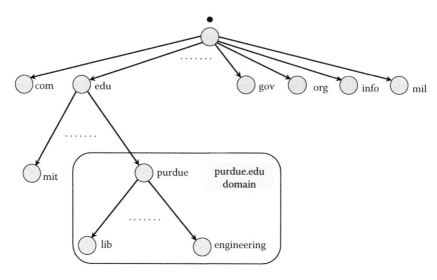

FIGURE 3.2
The hierarchical structure of domains in the Domain Name Service.

1. The computer requests the IP address of www.lib.purdue.edu from a local ISP's DNS Server.
2. The local ISP's DNS Server in India queries the root name server for the IP address. Recognizing that the requested address lies in the *edu* domain, the root name server responds by sending the DNS Server the IP address of the *edu* name server.
3. The DNS Server then queries the name server at the *edu* node, which responds by sending the address of the name server at the *purdue.edu* node.
4. The DNS Server then sends a request to the name server at *purdue.edu* node in response to which it receives the IP address of the name server at the *lib.purdue.edu* node.
5. The query from the DNS Server to the name server at the *lib.purdue.edu* node returns the IP address of the host corresponding to the www.lib.purdue.edu.

The actual implementation could involve some optimizations such as caching the addresses. It is also possible that the name server at *purdue.edu* could store the table for the entire purdue.edu domain and could return the IP address for www.lib.purdue.edu without referring the querying DNS Server to its descendant nodes. Such optimizations have been disregarded for conceptual simplicity.

In the DNS system the root node is a single point of failure. Therefore, the root name server is mirrored in 13 locations to provide redundancy. At the time of this writing there are 312 TLDs [TLDs 2012].

The advantage of such hierarchical distributed database is that changes at a node do not necessitate network-wide update. For example, the lib.purdue .edu domain is allocated a set of IP addresses that it is free to assign to the hosts under its control. If one of the host computers is decommissioned and the IP address reassigned to a different computer, then only the name server at lib.purdue.edu has to be updated. The rest of the network will receive the updated information whenever the name server at lib.purdue.edu is queried. Such delegation of naming privileges makes the DNS scalable.

The responsibility for naming the hosts and the subdomains is also delegated to the administrators responsible for the domains. The only restriction is that sibling nodes have distinct names.

Packet Switching

The Internet is based on two important paradigms—*connectionless packet switching* or just *packet switching* for short, and the *TCP/IP protocol*. The *packet switching paradigm* was proposed by Leonard Kleinrock in 1961 [Kleinrock 1961]. In order to describe the paradigm of packet switching it is useful to start by considering an alternative communication mode—*circuit switching*. In circuit switching mode, communication between two endpoints is enabled by dedicating a circuit—a communication channel—that connects the two endpoints for the entire duration of their communication. For example, a call between two telephones in the pre-Internet era involved committing a circuit to the call for the entire duration of the conversation. Committing a circuit ensures that the communication between the endpoints is not affected by the other traffic in the network and hence guarantees a certain degree of reliability. However, it could lead to a suboptimal use of network resources as the full bandwidth of dedicated circuits may not be used in the communications.

An alternative is the so-called connectionless packet switching, or *packet switching* for short. In packet switching mode, the digital data to be transmitted is partitioned into smaller pieces called *packets*. The packets are then independently transmitted from the source to the destination through the network. Each packet could follow a completely different path in its journey from the source to the destination. When the destination receives all of the packets it assembles them to reconstruct the original message.

The packet switching paradigm is clarified by considering a rather contrived analogy. Consider sending a book by postal mail in a manner that reflects the packet switching philosophy. One would then disassemble the book and mail each page in a separate postal envelope. All of these envelopes would be mailed from the source address to a common destination address. The destination address as well as the sender's address would have to be written on each envelope. Each of the envelopes could, in principle, traverse a different path. Upon receiving all the envelopes the recipient would reassemble the book.

The packet switching mode in digital communication bears considerable similarity to the contrived analogy just described. Each of the packets has to be prepared as an independent "envelope"—complete with both the sender's and the receiver's addresses. The IP protocol is used for addresses. Second, since the packets may traverse different paths to get to the destination, they may not reach the receiver in the same order in which they were sent. Therefore, the packets must carry sequencing information—like the page numbers—to enable the receiver to reassemble the packets in the correct order. Finally, some of the packets may not reach the destination due to network errors. Faced with such adverse events, the sender and receiver must be able to use some protocol to detect loss of packets and initiate retransmission.

The TCP/IP protocol specifies the guidelines for the segmentation of the original message into packets, the preparation of packets, their transmission over the Internet, their reassembly at the destination, and finally for ensuring the end-to-end reliability of transmission in the event of loss of packets. The TCP/IP is an example of a communication protocol suite. Communication protocol suites specify guidelines for communication at several layers— from the hardware layer over which communication occurs to the application layer that interfaces with the end user. The so-called OSI (Open Systems Interconnection) model provides an abstract framework for communication protocol suites. Therefore, we digress to discuss the OSI model before describing the TCP/IP protocol.

Open Systems Interconnection Model

The OSI model, developed by the International Standards Organization (ISO), partitions the interactions occurring over a communication channel between two communicating devices (nodes) into the following seven abstract layers [Dostalek and Cabelova 2006, Edwards and Bramante 2009]. Each layer may be viewed as providing an encapsulated service to the layer above.

Physical Layer (Layer 1): This layer is concerned with the electrical and physical details of the communication link between the two communicating nodes. It is concerned with such hardware details as the conversion of digital data into signals that are transmitted between the communicating devices. It hides the hardware details of the communication from the data link layer, which is allowed to assume that all the physical aspects of bit-level transmission are taken care of by the physical layer.

Data Link Layer (Layer 2): Assuming that Layer 1 provides the service of transmitting bits between the communicating nodes the data link layer focuses on ensuring that the errors in the transmission service (offered by Layer 1) are detected and corrected. As a result, Layers 1 and 2 appear to Layer 3 as a service that reliably transmits the binary data.

Network Layer (Layer 3): Layer 3 focuses on packaging data and routing it through the network, possibly in a multihop journey, from a source node to a destination node. Layer 3 relies on Layers 1 and 2 to handle reliable bit-level transmission in each hop of the multihop journey. The routers in the Internet operate in this layer. The IP (Internet protocol) specifies guidelines at this layer. Layers 1 to 3, taken together, provide a service, namely, the transportation of given data from the source node to the destination node.

Transport Layer (Layer 4): The transport layer interacts with the network layer in a manner that is analogous to the interaction of the data link layer with the physical layer. Recall that the data link layer focuses on ensuring the reliability of the bit-level communication occurring in the physical layer. Similarly, the transport layer assumes that the network layer provides the service of routing data from a node in one network to a node in a, possibly, different network (across a mesh of routers in the Internet). It focuses on ensuring the reliability of the service provided by the network layer. Thus, it is concerned with ensuring that data transmitted across the network has reached the destination. It is concerned with issues such as the flow control (to handle congestion). The TCP (Transmission Control Protocol) specifies guidelines at this layer.

Session Layer (Layer 5): This layer is responsible for establishing, managing, and terminating the connection between the two communicating processes—one at the source node and the other the destination node. For example, when a browser seeks to retrieve a web page from the web server on a remote computer the connection between the browser and the server are established at this layer. Once the connection is established it invokes the service offered by the transport layer and the layers below it to reliably transmit the data between the browser and the server.

Presentation Layer (Layer 6): At the sender's end, this layer prepares the data from the sender application (say, a web page being sent by a web server) for transmission across the network. For example, encryption of the transmitted data by the sender occurs in this layer.

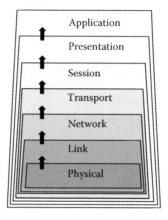

FIGURE 3.3
The seven layers in the OSI model. The arrows indicate that each layer can be viewed as providing an encapsulated service to the layer above. Protocols provide specifications for individual layers. For example, TCP is a transport layer protocol while IP is a network layer protocol.

At the recipient's end it converts the data received over the network (say, the received web page) into a form suitable for the application process (browser) that consumes the received data. For example, decryption of the received data would occur in the layer at the receiver's end.

Application Layer (Layer 7): This layer interacts directly with the application software serving as the window through which the application software accesses the network.

The seven layers are illustrated in Figure 3.3. The layered architecture of the OSI model permits a layer to interact only with the layers above and below it. Each layer operates over the service offered by the layer below it and, in turn, offers a service to the layer above it. The implementation details of the service in a layer are hidden from all other layers.

Structure of a Data Packet

As mentioned above the guidelines for preparation of data packets are specified at the network layer by the Internet Protocol. We take a closer look at the anatomy of a data packet in the Internet Protocol [RFC 791, 1981].

A data packet—also called an IP datagram—contains two types of information: (1) *a header comprising control information (fields A-H)*, and (2) *actual*

TABLE 3.1

Description of the Fields in the IP Data Packet

Field	Bits	Description
A	0–3	IP Version (e.g., IPv4 : 0100 in binary code)
B	4–7	Header length in 32-bit words (e.g., 0101 means 160-bit long header)
C	16–31	Total length of the datagram in bytes
D	64–71	Hop count field (number of remaining hops)
E	72–79	Protocol to be used to parse the data field (e.g., 00000100: TCP)
F	80–95	Header checksum (used for error-checking in header)
G	96–127	IP address of the source
H	128–159	IP address of the destination
I	160–	Data payload (length of which can be computed using fields B and C)

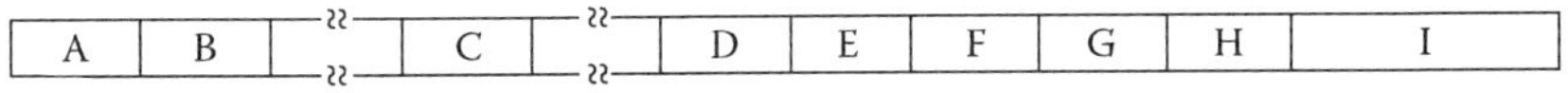

A	B		C		D	E	F	G	H	I

FIGURE 3.4
Format of an IP datagram.

data being transmitted (field I), called the *payload.* The fields in an IP packet are described in Table 3.1 and illustrated in Figure 3.4.

The fields G and H contain the source and the destination addresses, which are analogous to the from-address and to-address in postal communication. The IP datagram itself is analogous to a postal envelope, with the data payload being analogous to the contents of the envelope. The hop count field is discussed below. The data payload of the IP datagram is a TCP segment, which is prepared per the Transmission Control Protocol (TCP) described in the next section. The other fields are self-explanatory, and the reader is referred to Cerf and Kahn [1974] and Dostalek and Kabelova [2006] for a more detailed discussion of the fields in an IP datagram.

Transmission Control Protocol (TCP)

In packet switching the data to be transmitted is divided into smaller segments each of which is transmitted independently with no guarantee that all of the data packets will reach the destination. Individual data packets could get lost either because of glitches such as failure of routers or errors that might creep into the packet due to faulty transmission channels. In either case, the sender and the receiver must have a way of determining whether all of the transmitted data has been been received. If some of the transmitted data does not reach the receiver, then the lost data must be resent.

The *Transmission Control Protocol* (TCP) specifies the guidelines for dividing user data into smaller segments, packaging the segments, for setting up a communication between the sender and the receiver and for ensuring the reliability of the communication. A brief overview of these guidelines is presented below. For additional details, the reader is referred to [RFC 793, 1981].

TCP Segment: The TCP protocol provides guidelines for dividing data that is to be transmitted over the Internet into small chunks and packaging each chunk into a TCP segment. A TCP segment has a header that contains control information about the segment and a data payload. The header is usually about 20 bytes or 160 bits long (although it could be a little longer if one includes options). Figure 3.5 and Table 3.2 illustrate the anatomy of a TCP segment.

Packaging the IP Datagram: Figure 3.6 illustrates how the data get packaged into IP datagrams (packets). The data is first divided into several segments. Each segment is packaged into a TCP segment that includes a TCP header (described above) and the data segment. The TCP segment is then packaged into an IP datagram that includes an IP header (described above) and the TCP segment.

Handshake Protocol for Establishing a Connection: To ensure reliable communication between the sender and receiver a handshake

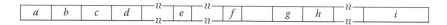

FIGURE 3.5
Format of a TCP segment.

TABLE 3.2

Description of the Fields in a TCP Segment

Field	Bits	Description
a	0–15	Port number at the source
b	16–31	Port number at the destination
c	32–63	Sequence number
d	64–95	Acknowledgment number
e	107	Acknowledgment bit
f	110	Synchronization bit
g	112–127	Size of the window that the receiver is willing to receive
h	128–143	Checksum (error-checking)
i	160–	Data payload

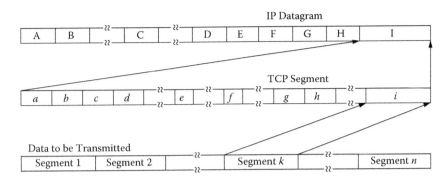

FIGURE 3.6
Illustration of the packaging of an IP datagram.

protocol is used to establish a connection in TCP/IP scheme [Dostalek and Kabelova 2006]. For example, if a user seeks to transmit some data to Purdue's web server then a connection is established between the user's computer and Purdue's web server using the following three steps.

Step 1: To initiate a connection with Purdue, the user transmits a packet in which the SYN bit (field *f* in the TCP segment) is set to 1, and the sequence number (field *c* in the TCP segment) is set to some random number, say *M.*

Step 2: Purdue acknowledges the request by transmitting back a packet in which the SYN and ACK bits (fields *e* and *f* in the TCP segment) are set to 1, and the acknowledgment number (field *d* in the TCP segment) is set to *M+1.* The sequence number (field *c* in the TCP segment) is set to another random number, say, *N.*

A packet in which the SYN and ACK bits are 1 and the acknowledgment number is *M+1* tells the user that the packet is a response to the connection request corresponding to the random number *M.* The random number *M,* which the user shares only with Purdue, enables the user to ignore spurious acknowledgment packets that are not responses to the request that the user sent Purdue. At this point the user's computer knows that the communication channels to and from Purdue are operative because he knows that the request he sent to Purdue and Purdue's acknowledgment have both been successfully transmitted. However, Purdue only knows that the channel from the user to Purdue is operative since all that it has received is a request from the user. There is no confirmation yet that its communication to the user has gone through.

Step 3: The user sends a packet to Purdue in which the SYN and ACK bits are set to 1, the sequence number (field c) is set to *M+1*, and the acknowledgment number (field d) is set to *N+1*. When Purdue receives this packet, it knows that the user has received its communication and hence the channel from Purdue to the user is operative.

Loss of Data Packets: Since packets are routed independently over the global network of routers a packet could get lost on the network for many reasons. For example, one of the reasons for the loss could be that the packet traverses a path that is longer than is allowed. Specifically, the hop count (field D in IP datagram; see Table 3.1) contains the number of remaining hops for the datagram. Each time the datagram hops from one router to another this count is decremented by 1. When the count reaches zero the next router concludes that the datagram has exceeded the allowed number of hops and discards the datagram.

Retransmission of Lost Data Packets: Observe that the user data is divided into different segments each of which is transmitted in a separate datagram as shown in Figure 3.6. The sender waits for a pre-determined amount of time to receive an acknowledgment from the receiver signaling the receipt of a data packet. If the sender does not receive acknowledgment within the time-out period, then it assumes that the datagram has been lost and retransmits the datagram. Fields c and d in the TCP segment are used by the sender and receiver, respectively, to communicate the sequence number of the datagram and the acknowledgment of receipt. The reader is referred to Dostalek and Kabelova [2006] for details.

Error Correction (Checksum): After a data packet reaches a recipient, the integrity of the data needs to be verified by the recipient. The checksum field contains bits that can be used to test the integrity of the data. The sender of a TCP segment uses the bits in the fields $a, b, c,$ $d, e, f, g,$ and i and a string of 16 zeros (bits) in the field h to compute a 16-bit number, which is then inserted into the field h in the TCP segment. The algorithm used to compute the checksum involves simple 1's complement arithmetic [RFC 793]. The TCP segment is then transmitted to the receiver. The receiver computes a 16-bit number using the same algorithm as the sender but using the checksum bits instead of zeros in field h. If the received data is the same as the sent data, then the output of the receiver's calculation should be a string of 16 1's. If the receiver obtains any other result, then it concludes that the transmitted data has errors in it, and the datagram is discarded.

The preceding paragraphs describe the structure of a TCP segment, how a TCP segment fits into an IP datagram, the handshake protocol a sender and receiver use to establish a connection, the hop count field that is used to limit the number of inter-router links an IP datagram traverses, the time-out mechanism used to detect loss of IP datagrams, and finally the error correction mechanism used to verify the integrity of the received data. The Transmission Control Protocol, described above, is part of a protocol stack called TCP/IP. For additional details about TCP/IP, including details about how a recipient reassembles IP datagrams, the reader is referred to Cerf and Kahn [1974] and Dostolek and Cabelova [2006].

The IP datagrams, described above, are transported between the end nodes of the Internet by a globally distributed network of routers, as shown in Figure 3.1. Internet's routers are owned by different networks and work cooperatively to provide a seamless global data transport service. The main functionality of an individual router is described in the next section.

Routers

A router is a hardware unit that is typically connected to several networks on the Internet. When it receives a data packet with a specified destination IP address the router consults its own routing table to identify the next router to which it must forward the packet, and transmits the packet to that router. The backbone of the Internet uses *core routers*, which can forward large volumes of data at high speeds. The routers closer to the periphery of the Internet, called the *edge routers*, are relatively less powerful.

A *routing table* typically contains a listing of reachable destination networks, the cost of sending a packet to each such destination network (in terms of the length of the path), and the router to which an incoming packet should be forwarded in order to route the packet to a target network. The information stored in a routing table could be either static or dynamic. One of the pitfalls associated with dynamic update of routing tables is the possibility of creating loops in which packets could get trapped. See Dostalek and Cabelova [2006] for a detailed discussion of routing.

Who Pays for the Internet?

The data flowing between a user's computer and a destination web server goes through a network of routers. In order to exchange data the two end-points of communication need to connect to the global network of routers. The *local Internet Service Providers* enable end users to connect to the global network of routers. The local ISPs charge their clients for providing connectivity to the Internet.

The network of routers is not owned by a single organization. Rather the ownership of the routers is globally distributed and organized into multiple tiers. At the top of the hierarchy is the collection of Tier 1 networks and the peering interconnections among them. In a pure peering interconnection between two networks both the networks agree to use each other's network resources without paying for them. Tier 1 networks are large backbone networks at the core of the Internet. They are characterized by large network sizes and high bandwidth communication links among its nodes. Tier 1 networks are distinguished in that they do not pay others for using the resources on the Internet. Below Tier 1 networks the classification gets somewhat fuzzy. Tier 2 networks are smaller networks that pay one or more Tier 1 networks to gain access to their network resources. Tier 2 networks could also have peering interconnections to other Tier 2 networks. In turn, the Tier 2 networks pass on the costs involved in gaining access to the Tier 1 networks to lower tier networks that seek network access through Tier 2 networks. Cascading down the hierarchy, one reaches the lowest tier comprising the local ISPs [Oppenheimer 2011]. Thus, in essence, the cost of operating the Internet is borne ultimately by the end users with the various network tiers providing the service of maintaining the network of routers.

The ISPs often reduce their expenses by setting up peering connections, called *Internet Exchange Points* (IXPs), with other ISPs within a geographic region. IXPs are physical network switches that interconnect the networks of the participating ISPs and are usually located close to the regions serviced by the participating ISPs.

Governance

There are two main recurring tasks that need to be performed to keep the Internet running: (1) *assigning* and *managing IP addresses* and *domain names* and (2) *stewarding the continuous evolution of the Internet.* Currently, a private nonprofit organization called the Internet Corporation for Assigned Names and Numbers (ICANN) coordinates the assignment of new IP addresses and the management of assigned IP addresses across the globe [www.icann.org]. Another international nonprofit body, called the Internet Engineering Task Force (IETF), is involved in developing new open standards and protocols for the Internet [www.ietf.org].

ICANN operates a unit called the Internet Assigned Numbers Authority (IANA), which is responsible for the coordination of IP addresses [www.iana .org]. The IP address allocation follows a hierarchical process. The IANA has

TABLE 3.3

The Regional Internet Registries and Their Associated Regions

Region	RIR
Africa	African Network Information Center (AfriNIC)
Asia/Pacific	Asia-Pacific Network Information Center (APNIC)
North America	American Registry for Internet Numbers (ARIN)
Latin America and some Caribbean Islands	Latin American and Caribbean Network Information Center (LACNIC)
Europe, Middle East, and Central Asia	Reseaux IP Europeens Network Coordination Center (RIPE NCC)

delegated address allocation authority in the five broad regions to Regional Internet Registries (RIRs) shown in Table 3.3 (see [RIR 2013]). The RIRs in turn allocate IP addresses to National Internet Registries (NIRs). The NIRs allocate addresses to Local Internet Registries (LIRs) and the LIRs allocate addresses to Internet Service Providers (ISPs), who assign addresses to end users. More information about IANA and its activities can be found at www.iana.org.

New standards for the Internet are developed by working groups operating under the umbrella of the IETF. Membership in these working groups is open to any interested individual and has no membership fee. The discussions in these groups are carried out mainly through mailing lists. A new standard developed by a working group is published as a document called *Request for Comment* (RFC). The standards are open. Further details about the IETF, its activities, and RFCs can be found at www.ietf.org.

History of the Internet

The following summary of the evolution of the Internet is based on an excellent account of the Internet's history written by some of the original architects of the Internet. The reader is referred to their article for further details [Leiner et al. 1997].

The first paper on packet switching was published in 1961 by Leonard Kleinrock [1961]. He also published the first book on packet switching in 1964. The following year the first recorded digital communication between computers was achieved by Lawrence Roberts and Thomas Merrill when they succeeded in transmitting data between a TX-2 computer in Massachusetts and a Q-32 computer in California, over a dial-up line. Their work heralded the dawn of computer networking. In 1966, Roberts moved to DARPA and began work on building a digital communications network, called the

ARPANET, when they succeeded in transmitting data that would link a distributed network of computers.

The diversity of computers and operating systems posed an immediate challenge to the design of ARPANET: how were two computers with different hardware and software environments to be interfaced in order to allow them to communicate? The problem was overcome with a far-sighted solution that, downstream, played a critical role in making the Internet architecture scalable. Wesley Clark suggested building gateway units—called *Interface Message Processors* (IMPs)—that would act as the interface between a local network of computers and the ARPANET. An IMP interfaced to the computers in its local area network, at one end, and to the ARPANET at the other end. The machine-dependent details of a computing system were filtered out by the IMP, allowing the ARPANET to interact with every IMP using a global protocol. The IMP embodied an important design feature: *it restricted the machine-dependent interfaces to the edge of the ARPANET allowing the core of the ARPANET to remain simple and machine-independent.*

The DARPA contract to build the IMPs was won by Bolt, Beranek, and Newman Inc. (BBN) in 1968. A BBN team headed by Frank Heart built four IMPs for DARPA. The first IMP was installed in Kleinrock's Network Measurement Center at the University of California–Los Angeles in 1969. It became the first ARPANET node. By the end of 1969 the other IMPs were installed at Stanford Research International (SRI), University of California–Santa Barbara, and University of Utah.

The second important design decision made in ARPANET was to *standardize the communication protocol for the host computers and the IMPs.* By the end of 1970, the Network Working Group developed the communication protocol—called the *Network Control Protocol* (NCP)—that governed the communications between host computers. The NCP assigned numeric addresses to computers that were connected to ARPANET, a practice that persists to this day. By imposing a universal communication protocol on all hosts and IMPs in the very early stages, the designers of ARPANET preempted fragmented evolution of communication protocols. Further, ARPANET adopted the packet switching communication mode from its early stages.

Unprecedented functionalities began to mushroom on the fledgling ARPANET almost immediately. By the end of 1970 it was possible to transfer files across the network. People could access data at remote locations without having to physically travel to the locations. By 1971 remote logins became possible. ARPANET implemented the *Terminal Interface Processor,* which enabled users to connect to the network from a terminal using a dial-up line. In 1971 Ray Tomlinson developed what was probably the most transformative application on ARPANET—*the e-mail.* Tomlinson was also responsible for introducing the @ symbol in electronic mail addresses.

Although ARPANET was evolving impressively it had a shortcoming. It was a single homogeneous network governed by a common set of protocols. The embryonic version of today's Internet began to take shape in 1973 when Robert Kahn, one of the members of the team that designed the IMPs at BBN, started pushing for an *open networking architecture* in which individual networks, conforming to different protocols, could be interconnected. Interconnecting disparate networks was called *internetting* and it ushered in the current avatar of the Internet. Kahn's initial efforts focused on internetting ARPANET with DARPA's packet radio network and packet satellite network.

Kahn's attempt to achieve internetting among ARPANET, the packet radio network, and the packet satellite network, exposed the limitations of the NCP. The NCP was designed to address the IMPs within ARPANET. It did not have the built-in capability to address resources in other networks. Further, the NCP was not concerned with the reliability of communication between the computers, which it assumed was the responsibility of the ARPANET infrastructure. While the assumption may have worked well within a single network such as ARPANET, it was no longer tenable when disparate networks were interconnected. There was a need for a communication protocol that not only accommodated interconnections among different networks but also ensured reliability of communication between end nodes. Both of these functionalities were built into a new *TCP/IP* communication protocol that Robert Kahn and Vint Cerf designed in 1973. *The development of the TCP/IP protocol enabled the internetting of disparate networks, and was a pivotal event in the evolution of the Internet.*

The principles that guided Kahn's work, articulated in Leiner et al. [1997], bear direct relevance to I-2 and are summarized below.

- *"The operation of the Internet must not have a global control. Rather control was to be distributed.*
- *No individual network, seeking to connect to the Internet, must be required to make internal changes in order to do so.*
- *Communication on the Internet was to be on a best-effort basis, with no guarantee of delivery of a data packet. Reliability was to be implemented through retransmission when necessary.*
- *The computers on the Internet would be assigned globally unique addresses."*

Kahn overtly sought to accommodate the diversity of networks, imposing few constraints on the individual networks that wanted to connect to the Internet. The globally distributed operational control encouraged organic evolution of the Internet. The 32-bit IPv4 addressing scheme that was adopted, however, underestimated the number of networks and computers that would connect to the Internet in the years ahead. Initially, the IPv4 address allocated 8 bits to label networks and the remaining 24 bits to label computers within a network. The initial partitioning of bits seems to have been based on the expectation that no more than 256 large networks would connect to the Internet. Around 1973 the Ethernet protocol for *Local Area Networks* (LAN) was developed at Xerox PARC. The Ethernet protocol triggered a rapid proliferation of LANs, making 8 bits vastly inadequate for numbering the networks that connected to the Internet.

Large time-sharing systems like Tenex and TOPS 20 computers hosted the initial implementations of TCP/IP. The TCP/IP implementation for desktop computers was developed by David Clark. It was shown to be interoperable with the other implementations of TCP/IP, making it possible for desktop computers to become end nodes of the Internet.

Although the TCP/IP protocol implementations had been developed for both the large computers as well as the desktops, propagating it among the community of Internet users was a challenge. One of the most widely used operating systems in the 1980s was UNIX. The UNIX operating system was rewritten at the University of California–Berkeley to incorporate TCP/IP. The research community, which was a large user base of UNIX, began using the UNIX containing the TCP/IP protocol, enabling TCP/IP to take root among early Internet users. The strategy of using a popular operating system for propagating a critical protocol played a pivotal role in facilitating the widespread adoption of the Internet. The adoption of TCP/IP proceeded swiftly. In 1980 TCP/IP became a defense standard. On January 1, 1983, ARPANET transitioned from NCP to TCP/IP.

In the early days of ARPANET, relatively few computers were connected to the network. A file called HOSTS.TXT maintained a table of computers connected to ARPANET and their numeric addresses. The master copy of HOSTS.TXT was maintained on a computer in SRI and was made available to all the computers connected to the network. As the size of the network grew, maintaining the addresses of all the computers that were connected to the Internet in a single table was no longer a scalable solution. There was a need for a distributed database that (*i*) maintained the mappings from computer names to their numeric addresses and (*ii*) could be queried by any computer on the Internet. Such a distributed database, called *Domain Name System* (DNS), was designed by Paul Mockapetris in 1983. It is a hierarchical database mapping the human-friendly address of a computer (such as www.lib.purdue.edu) to the IP address corresponding to it (such as 128.210.126.182). By querying the DNS using a human-friendly address as key, one can retrieve the IP address corresponding to the computer.

In the 1980s several specialized networks emerged. Prominent among them were MFENET, HEPNET, SPAN, CSNET, USENET, BITNET, JANET, and NSFNET. These noncommercial networks served special communities and had little motivation to be interoperable. Among these networks, NSFNET, which was supported by the U.S. National Science Foundation, was poised to play a critical role in the next stage of the Internet's evolution—making the Internet spill over from specialized communities, of mostly researchers, to the world population at large. The calculated strategic decisions that were made to steward the Internet through this transition are noteworthy.

NSFNET was intended to serve all members of the higher education community, regardless of their disciplines. In 1985 Dennis Jennings, who led the NSFNET program, helped make TCP/IP mandatory on NSFNET. Steve Wolff, who took over the leadership of NSFNET in 1986, was instrumental in setting the NSFNET on a path that would enable it to spin off a global network that was supported by nonfederal funds.

First, NSF partnered with DARPA to foster interoperability between NSFNET and the existing ARPANET. The federal agencies created the Federal Networking Council (FNC) that coordinated the shared use and management of the federally funded infrastructure, such as transoceanic circuits and the Federal Internet Exchanges. FNC cooperated with agencies in other continents, such as RARE in Europe, to extend Internet support to the global research community. Partnerships such as these helped interconnect the existing networks to forge a global infrastructure.

Second, NSF activated the following phased plan to privatize the expansion and maintenance of the Internet infrastructure.

1. The regional networks on NSFNET were permitted to seek commercial customers for its network facilities locally. This helped expose the commercial sector to the potential of the Internet.

2. At the same time that NSF permitted the local commercial use of its NSFNET resources, it prohibited the use of NSFNET infrastructure for commercial use at a national level. This dual strategy encouraged the commercial sector to build a parallel national infrastructure that could be used for commercial purposes, giving rise to such large-scale commercial networks as PSI and UUNET.

3. In 1990 ARPANET was shut down. In 1995 NSF funding for NSFNET's backbone was terminated. NSF distributed the leftover funds to the regional networks to enable them to buy connectivity to the national-scale private networks that had emerged, thereby providing early customers to the fledgling private infrastructures. NSF invested about $200 million into NSFNET between 1986 and 1995. By the time NSF finally discontinued funding for the NSFNET its strategic privatization plan had created a privately funded Internet infrastructure that sprawled across all the continents.

The most significant recent advance in the Internet technology is the birth of the mobile Internet. In 2008, the number of mobile broadband subscribers exceeded the number of fixed broadband subscribers [ITU 2009]. The mobile Internet is discussed in greater detail in Chapter 5.

Summary

The Internet, a successful distributed global infrastructure, has twofold relevance to I-2. It provides the bedrock on which the I-2 infrastructure will operate. Second, its history and architecture contain valuable lessons about building successful global infrastructures. Accordingly, we have presented an overview of the Internet's architecture, and also an account of its evolution during its early years. In Chapters 7 and 8 we revisit the history and architecture of the Internet to glean guidelines for building I-2. This chapter also provides the background for the discussion of the web and I-2's architecture, in Chapters 4 and 10.

4

World Wide Web

At its inception, the World Wide Web, or just the *web* for short, it was envisioned to be just a distributed library of interlinked *hypertext* documents. The documents in the library resided on the computers connected to the Internet. The hyperlinks, embedded within the hypertext documents, provided a user-friendly mechanism for navigating the universe of documents—suggestively called the *docuverse* [McKnight et al. 1991, Nelson 1980]. Tim Berners-Lee, the "father" of the web, designed and implemented the infrastructure by harnessing the data transport capabilities of the Internet.

Belying the modest scope envisioned at its birth, the web has grown to become one of the most transformative game-changing inventions. Its birth, during 1989–1991, triggered the steep rise in the Internet usage shown in Figure 1.6 [W3C 1999]. Over the last two decades the web has inseparably woven itself into the fabric of our everyday lives, meeting Mark Weiser's criterion for profound technologies [Weiser 1991].

The web is relevant to the I-2 infrastructure for two reasons. The web is a successful global platform whose underlying design principles embody important lessons about building large distributed infrastructures. Second, it provides the substrate framework for web-enabled services—the atomic units of transaction in I-2.

The following sections present a high-level overview of the web's architecture, its early history, and the design criteria used to architect the web. The web motivated the emergence of an architectural style called *REpresentational State Transfer*—REST for short—that has played an influential role in the evolution of the web's design. REST encapsulates several guidelines for designing scalable distributed infrastructures and is also playing an important role in the context of the so-called RESTful web services. An overview of REST, as well as further discussion of the web's design criteria, are deferred to Chapter 8.

Architecture of the World Wide Web

The architecture of the web, like that of the Internet, is easily understood by tracking a web-based transaction. Again, we will consider a simple

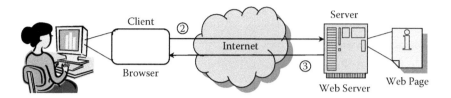

FIGURE 4.1
Schematic illustration of a web transaction.

hypothetical transaction—in fact, the same transaction that we considered in Chapter 3.

Recall the transaction from Chapter 3, shown in Figure 4.1, in which an end user requests the web page with the address *http://www.purdue.edu/index. html* through her browser. We reexamine the transaction from the perspective of the web.

- In the first step (not shown) the user provides the address of the web page—in this example, http://www.purdue.edu/index .html—to a web browser and instructs the browser to retrieve the web page.
- The browser in turn establishes a connection with the web server that hosts the web page and requests the server to send the web page. The communication between the browser and the remote web server is governed by the HyperText Transfer Protocol (HTTP) that was designed specifically for the transport of hypertext documents, such as web pages.
- The web server transmits the requested web page to the browser, again using the HyperText Transfer Protocol.
- In the final step (not shown) the browser translates the web page, written in the HyperText Markup Language (HTML), into a visual display of the web page on the screen for a human end user.

The unit of transaction in the above interaction is the web page, which is an instance of a general construct called *resource*. The browser plays the role of a client* that requests a resource—the web page—from a server, which in this case is the web server software. The client–server architecture style that underlies the above interaction between a browser and a web server is discussed in greater detail in Chapter 8. In the following sections, we take a closer look at the details of the above transaction.

---

* The interaction is discussed in greater detail in Chapter 8. The browser is actually a user agent, and the term *client* is reserved for a stub that the browser invokes. With slight abuse of notation, we call the browser the *client* in this chapter.

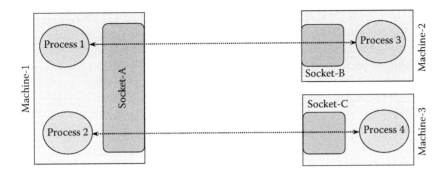

FIGURE 4.2
Use of sockets in network communications.

Ports and Sockets

The notion of *port* plays an important role in HTTP transactions. A port is a virtual "tray" on a device that is used to field incoming data. Ports can be addressed by other devices in the network. Ports on a machine are typically assigned numbers ranging from 0 to 65535 (16 bits) with each port associated with a unique communication protocol. For example, port 80 is usually earmarked for communications that use HTTP protocol, port 443 for communications that use the HTTPS protocol, port 25 for communications using SMTP protocol, and port 110 for communications using the POP protocol, to mention a few examples [Centos 2013].

The concatenation of the IP address of a machine and a port number gives a construct called the *socket* [Winett 1971]. An example of a socket address is http://www.purdue.edu:80/, where http://www.purdue.edu is the IP address and 80 the port number at that IP address. Sockets serve as the endpoints for network communications.

Multiple processes, such as Process 1 and Process 2 in Figure 4.2, could use the same socket for network communication. Communications through a common socket are resolved using socket pairs. The dotted lines indicate the active connections between processes running on different machines. The communication between processes 1 and 3 corresponds to socket pair (A,B), whereas that between processes 2 and 4 corresponds to socket pair (A,C). The incoming data at Socket A is routed to Process 1 or 2, depending on whether the socket address of the sender is B or C.

HyperText Transfer Protocol (HTTP)

The HyperText Transfer Protocol, or HTTP for short, is a protocol used in client–server communication on the web. For example, when a client (browser) wants to retrieve a web page from a server it sends a HTTP request to the machine on which the server runs. HTTP requests are usually routed to port 80 of the server's machine. On receiving the request the server processes it and returns its response to the client, again using HTTP. The toy example of client–server communication, presented below, serves to illustrate the protocol. The reader is referred to [Fielding et al. 1999] for a more detailed discussion of HTTP.

A HTTP request has three parts:

- A command that specifies the request
- A series of header lines that provide information about the request or about the contents of the body of the request
- Body of the request, which could contain data pertinent to the request.

For example, consider the following HTTP request.

> **GET** /path/resource HTTP/1.1
>
> **From:** abc@def.com
>
> (Body of the message)

The GET command tells the server that the client is requesting the resource at the specified path on the server's machine, and that the protocol being used in the request is HTTP version 1.1. The header line tells the server that the client's email address is abc@def.com. Optional information about the request could be communicated in the body of the message. The keywords of the HTTP protocol are in boldfaced characters. For a list of permitted commands see [Marshall 1997].

A HTTP response also has three parts.

- The status of the request
- A series of header lines that provide information about the response and/or about the contents of the body of the response, and
- Body of the response.

For example, in response to the above request, the server could send the following response.

> HTTP/1.1 **200 OK**
>
> **Content type:** text/html
>
> **Content length:** 1000
>
> <html> ...... </html>

The first line of the response confirms the HTTP version number followed by a numeric code of the request status and a verbal description of the status. Code 200 indicates successful processing of the request, summarized verbally by OK. The next two header lines specify the type of data being returned in the response body—a html file in this example—and the length of the body in bytes—1000 bytes. The body of the response contains the html file that begins with the tag <html> and ends with the tag </html>. For a more detailed discussion of HTTP the reader is referred to [Fielding et al. 1999].

Resource

The unit of transaction on the web is called a resource. A *resource* is defined to be any entity that has an identity [RFC 2396]. For example, a file on a server is a resource. In the early days of the web, the term resource was synonymous with entities such as files, images, movies, and programs that were stored in electronic format. Over the years, however, the term *resource* has evolved to encompass all entities that have identities. For example, an image stored on a server, the color red, a service, a toll booth operator, or a coffee mug are all examples of resources, because they have identities.

The expanded definition of a resource introduces new challenges to HTTP. Not all resources are retrievable over a network. For example, although a coffee mug is a valid resource a HTTP request cannot retrieve it. There was considerable debate about distinguishing between information resources such as documents and images, which are retrievable over the Internet, and noninformation resources, such as a physical object, which are not retrievable over the Internet.

For a while, a hash mark notation was used to describe a noninformation resource, while a slash notation was used for retrievable information resources. Thus, for example,

http-address/information-resource-name

format was used to refer to an information resource (such as a file or an image) while

http-address#non-information-resource-name

format was being used to refer to resources that could not be retrieved over the network. The "hash" versus "slash" debate appears to have ended. The distinction between information and noninformation resources is no longer communicated using # and /. Rather, the response to a HTTP GET command is used to distinguish between information and noninformation resources in HTTP communications. If a requested resource is a network-retrievable information resource, then the response to the HTTP GET request returns status code 200. Otherwise, the response returns a status code 303, indicating that a representation of the requested resource is not available at the server. For a history of the usage of this term see [Berners-Lee 2009]. The meaning of the term resource is further clarified in [RFC 2396] and [RFC 3986].

The main benefit of the notion of resource is that it provides an umbrella construct that subsumes the diversity of digital objects flowing across the web. The atom of transaction on the web is thus a uniform entity—*resource*. The architecture of the web is greatly simplified by viewing the web as a framework for transporting resources. The processing related to the internal details of resources, such as conversion of resources to desired formats or the rendering of resources, is restricted to occur at the edge of the web, that is, inside the clients and servers. The significance of the encapsulation provided by the resource construct is discussed further in later chapters.

Browser

A browser is a software service that provides an end user with a friendly interface to the World Wide Web. In the preceding example, the browser (1) accepts the address of the web page, (2) establishes a connection with the target web server that hosts the desired web page, (3) retrieves the web page from the server using HTTP, and finally (4) displays the retrieved web page in a human-friendly visual format on the user's screen.

The retrieval of a web page from a server by a browser involves the exchange of data packets over the Internet. Such exchange of data packets is governed by the TCP/IP protocol. The HTTP protocol operates on top of the TCP/IP protocol. An analogy would be the act of sending a payment for a utility bill by mail. The protocol for paying a utility bill, which is analogous to the HTTP protocol, involves enclosing a check made out to a utility company together with the information about the account toward which the payment is being made. The postal protocol for mailing the payment, which is like the TCP/IP protocol, involves enclosing all the documents in an addressed envelope, affixing stamps of the right value and mailing the envelope. Thus, the bill payment process uses the postal protocol for the physical transmission of the envelope and the bill payment protocol for preparing the contents of the envelope. In that sense, the bill payment protocol operates on top of the postal service protocol. Analogously, the web-based communications use the TCP/IP protocol for the transmission of data and the HTTP for higher-level exchange of resources.

Web Server

A web server is also a software service. It enables a document to be exposed on the web. A web server's primary function is to receive requests from clients (web browsers) and return the requested resources to the clients. In the above hypothetical transaction the web server at the address *www.purdue .edu* fields the request for the web page *index.html* from the browser (client) and responds by sending the requested web page to the browser. The file *index.html* residing on Purdue's computer would not be visible, or retrievable on the web, even if the computer is connected to the Internet, unless the computer runs the web server software service and includes the file in the basket of resources exposed by the web server service. Modern web servers perform many other functions besides the simple task of provisioning resources. An interested reader is referred to [Yeager and McGrath 1996] for additional details.

Search Engine

An important indexing service operating on top of the web is a search engine. A search engine performs two continuous activities: *update* and *index*. The web of interlinked documents is changing at a rapid pace. New resources are being added to the web even as existing resources are being updated or deleted. A search engine periodically interrogates the web to update its own knowledge of the web's contents. Subsequently, it semantically indexes the web's contents in order to enable end users to search the web rapidly. A user, querying a search engine using key words, is provided pointers to contents on the web that semantically match the key words.

Governance

The evolution of the World Wide Web's architecture is being stewarded by an international organization called the *W3C (World Wide Web Consortium)*. The development of new standards and protocols for the web occurs in working groups inside W3C. The working groups include W3C members and invited experts. Membership in W3C carries an annual fee. The end product of a working group's deliberations is a W3C Recommendation. Further details about the W3C, its activities, and recommendations can be found at www.w3c.org.

History of the Web

The following account of the birth and the subsequent development of the web is largely based on the account of its history written by the person who invented it. The reader is encouraged to read his article [Berners-Lee 1996], which contains a more detailed narrative on the early years of the web. Our motivation for discussing the web's history is to glean, from the details of its birthing process, lessons about incubating global infrastructures. Accordingly, our interest and the following account of the web's history are restricted to the birth of the web and its evolution in its early years.

The earliest glimmer of the web was in the oN-Line System (*NLS*) designed by Doug Engelbart, who is also credited with the invention of the computer mouse. NLS, which was developed in the 1960s, was an online system that used *hyperlinks* to enable users to browse through the stored documents.

Unlike the Internet, which had nebulous beginnings with contributions from many researchers, the beginnings of the web can be traced to a definite event in 1989 and to the vision of one person, Timothy Berners-Lee. In March 1989 Berners-Lee wrote a proposal to build a database of interlinked hypertext documents.*

Implementing the prototype of the web involved developing (1) a web server; (2) a web browser; (3) a protocol, operating on top of the Internet data transmission protocol, for the transfer of web documents; and (4) a markup language used to annotate hypertext documents. The annotation is intended to provide guidelines to a web browser about displaying the material in the hypertext document. In 1990 Berners-Lee built the first web server (*http://info. cern.ch* running on a NEXT computer) and the first web browser (which was called the *WorldWideWeb*). He also designed the *HyperText Transfer Protocol* (HTTP) that governs the transfer of the web's documents and the *HyperText Markup Language* (HTML) to annotate the documents for the browser's consumption. A portable browser that could run on any platform was developed in 1991 by Nicola Pellow.

A key decision that Berners-Lee made was to make the web technology freely available to the world community. The decision facilitated the worldwide adoption of the web in the years that followed.

The next significant event in the history of the web was the development of the Mosaic browser by Marc Andreessen and Eric Bina in 1993. Mosaic was the first browser that could display images. The following year search engines Lycos, Web Crawler, and Netscape Navigator were released. Audio streaming started in 1995. From about 26 public web sites in 1992, the web grew to about 250,000 web sites by 1996. The *Google* search engine was

---

* A *hyperlink* is a pointer to a resource that could be either within or outside the document containing the hyperlink. A document that contains both ordinary text and embedded hyperlinks is called a *hypertext document*.

launched in 1998. Between 1996 and 1998 the number of web sites increased from about a quarter million to about three quarters of a million.

In the first version of the web, called *Web 1.0*, the web was asymmetric. Users visiting a web site were largely consumers of the content published by the owners of the web site. The pervasive emergence of web-based platforms on which the users were allowed to be both the producers and consumers of content marks the next avatar of the web, termed *Web 2.0*. The poster child of Web 2.0 is the *Wikipedia* database, with *Facebook* and *Youtube* providing other noteworthy examples.

Design Criteria

As mentioned before, the web's design embodies lessons for architecting successful global infrastructures. Accordingly, in the following paragraphs we review the design criteria that Berners-Lee espoused [Berners-Lee 1996]. Among the many criteria he used, the following four are particularly critical for the successful emergence of the web as a scalable global infrastructure. The criteria were (in Berners-Lee's own words):

1. *"If two sets of users started to use the system independently, to make a link from one system to another should be an incremental effort, not requiring unscalable operations such as the merging of link databases.*

2. *Any attempt to constrain the users as a whole to the use of particular languages or operating systems was always doomed to failure.*

3. *Information must be available on all platforms, including future ones.*

4. *Any attempt to constrain the mental model users have of data into a given pattern was always doomed to failure."*

The design criteria sought to make the web an enabling platform that could adapt itself to interoperate with the existing data items. From the beginning, the web operated as parallel technology to existing systems and did not *coerce* the end users to change their operating system, language, or data model in order to connect to the web. Thus, the web allowed for conventional text documents to coexist with the hypertext documents, making no distinction between the two types. Such a *Non-Coercive Architecture* lowered the barrier for connecting to the web, and Berners-Lee observes that the principle of placing minimal constraints on the end users was a major factor in the web's widespread adoption.

Second, the design criteria ensured that the effort involved in growing the web was proportional to the magnitude of the changes and not to the current size of the web. Without this design feature, the marginal effort to grow the web would have escalated rapidly.

The third design criterion above separated the implementation of the web, which was expected to evolve, from the underlying architecture of the web, which was expected to endure without changes. This design criterion made it possible to upgrade, even change, parts of the web while preserving the underlying architecture. Building such flexibility forced the implementation of the web to adhere to modularity and information hiding. For example, the reference to a resource on the web does not contain any information that would prohibit the owner of the resource from replacing it with a different version of the same resource or even a different resource altogether.

Faced with a trade-off between keeping the web consistent and lowering the barrier for growth the web's design favored ease of growth. Specifically, keeping the web consistent requires that changes to the state of the web be propagated throughout the web. For example, if a document is removed from the web, then consistency requires the deletion of all the links pointing to the document as well or the web would be left with dangling links pointing to nonexistent resources. However, the consistency requirement would require the maintenance of a database of all the links pointing to resources. That is, each time a new link is added in the web the database would have to be updated, making the creation of new links a cumbersome process.

In the web's architecture the consistency requirement was compromised to preserve the ease of making changes. Thus, a document on the web can be deleted without updating all the references to the document. This feature enables users to create links pointing to a document without informing anyone, including the document's owner, about the new link. Conversely, a document's owner is allowed to remove the document without informing the users that have links pointing to it. Such operations make the web's state perpetually inconsistent. The inconsistencies are resolved over time by self-correcting mechanisms operating on the web.

Summary

As with the Internet, the web also has twofold relevance to I-2. It plays an essential role in the I-2 infrastructure. Second, its history embodies lessons about building successful global infrastructures. We have presented a coarse-grained account of the web's architecture, a summary of its history, and some of its design criteria. We revisit the web's architecture, history, and design criteria in Chapter 8. An understanding of the web's architecture will be assumed in the discussion on web services in Chapter 9.

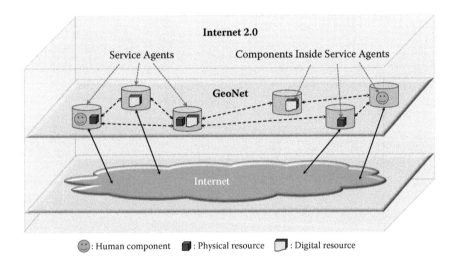

COLOR FIGURE 10.5
The Core Design Principle that sets the resolution scale for I-2.

5

The Mobile Internet and the Mobile Web

A stationary computer typically connects to the Internet through a home/office gateway and the Internet service provider's (ISP's) router as shown in Figure 5.1. The connection between the computer and the gateway, and between the gateway and the ISP router are typically fixed-line links. For example, the computer could be connected to the gateway through an Ethernet cable, and the gateway to the ISP router through a telephone line or broadband cable. Such wired connections greatly limit the mobility of the computer.

In contrast laptops and handheld devices such as smart phones are not typically tethered to other hardware. Rather, they communicate with the Internet through wireless channels. The increased mobility made possible by wireless links presents the challenge of sustaining the mobile device's connectivity to the Internet as it wanders both within the coverage area of the home gateway and across it into the coverage area of foreign gateways. Responding to the need to connect a growing number of mobile devices to the Internet, a new technology—the mobile Internet—has emerged. In this chapter we discuss the details of the mobile Internet.

The objects that I-2 seeks to connect to the cyber infrastructure are like the mobile devices in that they are usually not tethered. Therefore, the communication architecture in I-2 faces many of the same challenges that have been overcome in building the mobile Internet. Aspects of the discussion in this chapter will resurface when we discuss I-2 in later chapters.

The discussion of the mobile Internet has been divided into four parts in this chapter. In the first part, we will look at the wireless technologies used to connect mobile devices to hubs such as home/office gateways and wireless hotspots as shown in Figure 5.2.

In the second part, we will look at the protocols used to sustain connectivity to the Internet as the mobile device wanders outside the range of the home wireless blanket into a foreign wireless blanket.

An exciting paradigm for networking mobile devices is the so-called Mobile Ad-hoc Network (MANET) illustrated in Figure 5.3. Consider a collection of wireless-capable devices labeled A–G. Each of the devices has a finite communication range. For example, devices A and F are too far apart to be able to communicate with each other directly. However, devices B and G are sufficiently close to A that A can communicate directly with B and with G over wireless communication links. We say that the communication link between A and B, and between A and G, are active. If A and B move out of

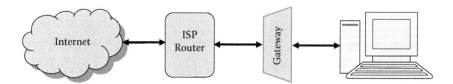

FIGURE 5.1
Fixed-line link between a stationary computer and the Internet.

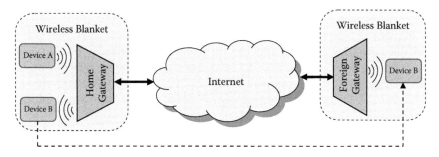

FIGURE 5.2
Device migration across wireless blankets.

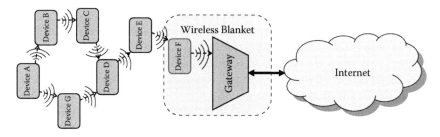

FIGURE 5.3
Mobile ad hoc network.

each other's communication range,* then the wireless link between A and B is dynamically broken, and we say that the link is inactive. At any instant of time a MANET is a network formed by the devices and the active communication links among the devices.

The nodes in a MANET could serve as routers that support the flow of data across the network. For example, if A seeks to communicate with F, it can do so even though A and F are not within each other's range. A can route the data to F through, say, the path A → G → D → E → F. In such a communication, the

---

* Since A and B could have different communication ranges, it might be possible for one node to send messages to the other while the other cannot. The network structure of a MANET is thus that of a directed graph. In this chapter, we ignore such details and assume that all devices have the same communication range, and that the MANET has the structure of an undirected graph.

intermediate nodes G, D, and E function as "routers" forwarding the incoming data to the next node along the path. Since F is within the wireless blanket of a gateway to the Internet, A could gain connectivity to the Internet, using the bidirectional connectivity to F, over the MANET. Thus, a MANET can be used to extend Internet connectivity to those nodes that are outside the wireless blanket of a gateway, as shown in Figure 5.3.

Besides providing Internet access to distant nodes, a MANET also enables the devices to share data and intelligence among themselves giving rise to nontrivial cooperative behavior. The VANET, discussed in Chapter 2, is an example of such cooperative behavior. In the third part of our discussion in this chapter, we will focus on the mobile ad hoc network paradigm. We will continue the discussion on mobile ad hoc networks in Chapter 11.

Finally, we discuss the notion of mobile web. The bandwidth and display restrictions in mobile devices necessitate changes in the web contents to make them suitable for mobile devices. The new mobile web technology seeks to make the web more friendly to mobile devices.

Terminology

Wireless networks can be classified by their range. Thus, a *Wireless Personal Area Network (WPAN)* has a range of a few meters and is used to connect the personal wireless devices such as wireless printers, pointing devices, keyboards, headsets, and personal digital assistants. In contrast, the *Wireless Local Area Network (WLAN)* is used to interconnect devices within a range of a few hundred meters, and is characterized by high data transfer rate. A *Wireless Wide Area Network (WWAN)* has a range of up to tens of kilometers and is used to provide wireless coverage over citywide geographic regions [Rackley 2007].

Wireless networks can also be classified by their interconnection topologies. We mention two topologies as examples. In the star topology, shown in Figure 5.4, the client devices are connected to the central hub. They do not, however, communicate directly with each other. In the *peer-to-peer topology* all the nodes are on equal footing and are connected by communication links to the nearest neighbors.

Wireless Communication Protocols

For two nodes to be able to communicate over a wireless channel they need to have a shared understanding of the specifications of the communication channel at all the seven layers of the OSI model (discussed in Chapter 3). For

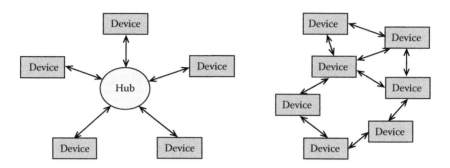

FIGURE 5.4
Star topology (left) and peer-to-peer topology (right) for communication networks.

example, the two nodes must be aware of the frequency they need to use in the transmission, the transmission speed (in bits/second), and the acknowledgment and synchronization rules used in managing the connection at the hardware level (Layer 1). The devices must be aware of the error detection and correction mechanisms used to ensure the reliability of the bit-level communication (Layer 2), and so on. A formal specification of the rules used for communication at all the seven layers is called the *communication protocol suite,* sometimes also called the *protocol stack.* For example, *ZigBee* is a wireless protocol suite used in WPAN. The specifications of a protocol suite at a single layer constitute a *protocol.* Thus, the IP (Internet Protocol) specifies the rules at the network layer of the TCP/IP stack, while the TCP (Transfer Control Protocol) specifies the rules at the transport layer of the TCP/IP stack. When a communication protocol is approved by a standards organization, such as ISO (International Standards Organization), it becomes a *standard protocol.*

Standard protocols play a critical role in promoting interoperability of devices. Two devices made by the same or different manufacturers can communicate with each other if they are designed based on the same communication protocol. Not surprisingly, a wide selection of communication protocols have been devised. Some of them are *proprietary protocols*—that is, protocols owned by an individual or a company—while others are *open protocols,* which are available for use by the community at large at no cost. We present a few widely used communication protocols below. The discussion is not intended to be comprehensive. Rather, the intent is to present some illustrative examples. The discussion is organized by the scale of the network for which the protocols are designed.

WPAN Protocols

The *IEEE 802.15 Working Group* has developed three communication protocols for the physical layer of WPAN [Ergen 2004]. They are the IEEE 802.15.3

protocol, designed for high bandwidth devices, such as those that transmit/ receive video data. The IEEE 802.15.1 protocol, on the other hand, was designed for medium bandwidth devices, and they consume correspondingly less power. For example, the 802.15.1 protocol is used by devices that transmit audio data. Finally, the IEEE 802.15.4 protocol is intended for devices that do not require high bandwidth, but rather need to consume low amounts of power.

The *ZigBee* protocol stack is built on top of 802.15.4, by complementing the protocol for the physical layer provided by 802.15.4 with the protocols for the upper layers of the OSI model. ZigBee operates at the global unlicensed frequency of 2.4 GHz, having a bandwidth in the range of 20-900 kilobits per second (kbps). Within the United States it also operates at 915 MHz and at 868 MHz in Europe, with the bandwidths being lower at lower frequencies. ZigBee consumes low amounts of power. Therefore, it is suitable for devices that do not have high bandwidth requirements but face severe constraints on the available power. ZigBee typically has a range of about 10 to 75 meters, making it well suited for home automation applications [Ergen 2004].

The TCP/IP stack, whose implementation requires considerable amount of power, was generally considered unsuitable for low-power devices, such as those targeted by the 802.15.4 protocol. On the other hand, the TCP/IP stack is widely used in Internet communications, and it was considered desirable to use the same protocol suite in the WPAN networks as well, to increase seamless interoperability between WPANs and the Internet. The *6LoWPAN* protocol suite—which stands for *IPv6 over Low Power WPAN*—was designed to operate on top of 802.15.4 standard protocol. It enables low-power devices to use TCP/ IP in their communications. 6LoWPAN achieves the reduction in the TCP/IP overheads and hence the power consumption by using an adaptation layer between the link and network layers of the TCP/IP stack [6LoWPAN 2009].

While ZigBee and 6LoWPAN are low-bandwidth, low-power protocol stacks built on 802.15.4, the *BlueTooth* protocol suite is built on top of the medium bandwidth 802.15.1 protocol. BlueTooth operates in the 2.4 GHz to 2.485 GHz frequency spectrum and has a range of up to 100 meters [Bluetooth 2012].

WLAN Protocols

The *IEEE 802.11* standard is a family of wireless communication protocols at the physical layer that is used for WLAN [Rackley 2007]. Members of the family operate at 2.4 GHz, 3.6 GHz, and 5 GHz. For example, the 802.11b protocol operates at 2.4 GHz, supporting data transmission rates of up to 11 megabits/second and has a range of a few hundred feet. On the other hand, the 802.11g protocol also operates at 2.4 GHz, supporting transmission rates of up to 54 megabits/second and also has a range of a few hundred feet. The 802.11n protocol operates at either 2.4 GHz or 5 GHz, supporting data transmission rates of up to 150 megabits/second and has a range that is nearly twice that of the 802.11b or 802.11g.

Wi-Fi is a communication protocol suite for WLAN operating on top of the 802.11 family of standards. It offers a high data transfer rate and has a range that is suitable for wireless connectivity to Internet through access points at home and office. Wi-Fi is also the popular choice for providing access to the Internet at public *hotspots*—wireless Internet access points—such as those in airports and libraries.

WWAN Protocols

IEEE 802.16 is a family of protocols at the physical layer intended for high data transmission rate and large-range communication [Rackley 2007]. For example, the 802.16m-2011 supports data transmission rate of up to 1 gigabits/ second and a range of up to 50 kilometers. The protocols in the family operate over a frequency range of 2 gigahertz to 66 gigahertz.

WiMAX, an acronym for Worldwide Interoperability for Microwave Access, is a protocol suite built on top of selected protocols in the 802.16 family and is designed for broadband wireless communication. It is used for wireless communications in WWAN over citywide geographic regions [Rackley 2007].

IP Masquerading

The mobile devices that connect to the Internet are not required to have globally unique IP addresses attached to them. Gateways use *IP Masquerading* to enable devices, including those that are not IP-enabled, to connect to the Internet, as shown in Figure 5.5. The network address translation plays a key role in supporting a mobile device's interaction with the Internet. We discussed the network address translation briefly in Chapter 3. We take a closer look at the service below.

Consider two mobile devices, say Laptops 1 and 2, that lack IP addresses and seek to connect to the Internet through a gateway that has an IP address. The devices are assigned internal IP addresses by the gateway. In this example, the Laptop 1 is assigned the IP address 192.168.1.100, while the Laptop 2 the IP address 192.168.1.105. The internal IP addresses (which typically are of the form 192.168.X.Y) are not globally unique, and are hidden behind the gateway. We'll take the IPv4 address of the gateway to be A.B.C.D, where A, B, C, and D are numbers in the range 0–255. In contrast to the internal IP

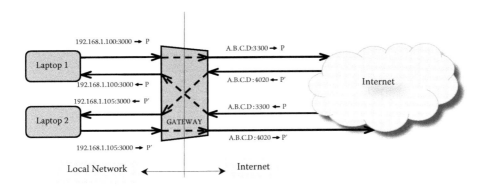

FIGURE 5.5
Illustration of network address translation and IP masquerading.

addresses of the laptops, the IPv4 address of the gateway, namely, A.B.C.D, is a globally unique IP address. External nodes on the Internet can use the address A.B.C.D to route messages to the gateway.

The two laptops could have several processes that access the Internet. Assume that a browser, say, Firefox, running on the Laptop 1 seeks to retrieve a web page on a remote server named P on the Internet. Simultaneously, a browser, say, Safari, running on Laptop 2 could also seek to retrieve a web page from the same or possibly different web server on the Internet, that we will call P'. When the Firefox browser on Laptop 1 sends a request to the remote server P, Laptop 1 assigns a port number, 3000 in this example, to the conversation between the Firefox browser and P. The message originating from the browser is dressed with a from-address 192.168.1.100:3000, and sent to the gateway. The gateway in turn assigns its own port number to the conversation between the Firefox browser on Laptop 1 and the web server P. If the port number used by Laptop 1, namely, 3000, is available it is used for the communication. Otherwise, it uses an available port number, which we have taken to be 3300 in this example. The gateway maintains the mapping from 192.168.1.100:3000 to its port 3300. When the gateway receives a message with a destination address A.B.C.D:3300, it knows that the message is intended for the local address 192.168.1.100:3000. The message is routed to Laptop 1, which in turn recognizes that its port 3000 is earmarked for the Firefox request and routes the response to the Firefox browser. The request-response communication pertaining to the Safari browser in Laptop 2 is handled similarly.

The Network and Port Address Translation occurring inside the gateway makes it possible for a large number of devices in a local network to connect to the Internet by sharing the single IP address of the gateway (A.B.C.D in the above example). The local address and port number such as 192.168.1.100:3000 is provided an identity on the Internet by mapping it to the address and port number A.B.C.D:3300, that has a unique identity on the Internet. The network address translation makes it possible to masquerade an entire local network behind the single IP address of the gateway. Especially when the devices are resource-constrained, like the objects connecting to I-2, it is desirable to let a single gateway handle the software overheads of connecting to the Internet using the heavy TCP/IP protocol suite.

Mobile IP Address

Mobile Internet enables devices to wirelessly connect to the Internet through gateways at different geographic locations. Ignoring the service agreements that may be required to use foreign gateways, we look at the two paradigms that a roaming mobile device can use to connect to the Internet. The first paradigm is the IP masquerading described above, in which a mobile device uses the IP address of the gateway to connect to the Internet. Some software licenses, however, are tied to the IP address of the device. In such instances the device needs to have a fixed IP address that travels with it as it accesses the Internet through different gateways. The *mobile IP* provides devices the capability to carry IP addresses with them as they roam. The following discussion of mobile IP is based on [Cisco 2001].

We revisit Figure 5.2 in which a mobile device is shown roaming from the wireless blanket of its home gateway into the range of a foreign gateway. Mobile IP implementations have several variants. In the following text, we present a scenario that illustrates the paradigm in a simple setting. For further details about mobile IP the reader can consult [Cisco 2001].

The first task that a device must perform when it roams into the wireless blanket of a foreign gateway is to discover the blanket. The discovery of the network could occur in one of two ways. The foreign gateway might advertise its network periodically. A mobile device fitted with a wireless network adapter listens for such advertisements and is able to discover the network. Alternatively, the device could announce its presence by sending a solicitation. The foreign gateway responds to the solicitation by broadcasting an advertisement.

Once the foreign network has been discovered, the device routes a message to its home gateway to set up the so-called *reverse tunnel* between the home gateway and the foreign gateway. Setting up a reverse tunnel involves verification of a service agreement between the home and foreign gateways and authentication of the request by the home gateway. Messages coming into the home gateway and intended for the mobile device are rerouted by the home gateway to the mobile device through the foreign gateway. Conversely, the messages from the mobile device to a remote node on the Internet are routed by the foreign gateway through the home gateway as shown in Figure 5.6.

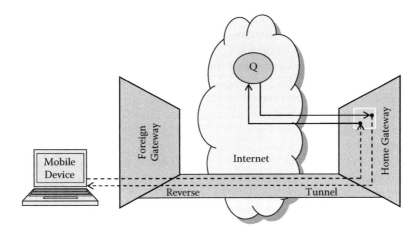

FIGURE 5.6
Mobile IP via reverse tunneling.

A remote node in the Internet—such as Q in Figure 5.6—communicating with the mobile device does not see that the mobile device has roamed outside its home gateway's range. The reverse tunnel between the two gateways also uses the Internet infrastructure.

Mobile Ad Hoc Network

A gateway provides an access point for connection to the Internet. A device that is within the wireless range of a gateway can connect to the Internet through the gateway. Oftentimes, however, many of the devices are outside the wireless blanket of the gateways that they can use. The *Mobile Ad hoc NETworks* (MANETs) provide an architecture that makes it possible to extend the wireless blanket of a gateway. Figure 5.3 and the discussion tied to it describe the basic notions about MANETs.

In a MANET, the wireless devices form an ad hoc network using each participating device as a router. The MANET increases the power consumption overheads on the individual devices since they participate in routing the messages of the other devices. However, the cooperative behavior provides Internet connectivity to distant devices.

MANETs are of interest not only for extending the coverage of wireless blankets but also for nontrivial sensing operations. The VANET described in Chapter 2 is an example of how MANETs can facilitate self-organization. The vehicles participating in the VANET are able to dynamically reroute themselves to avoid traffic congestion.

In the following discussion we will take a closer look at how a collection of autonomous wireless devices can organize themselves into MANETs. MANETs are the focus of ongoing research and provide the basis for an I-2 prototype proposed in Chapter 11. In the following paragraphs we describe how MANETs form and self-organize, using the concrete protocol suite ZigBee. The discussion is based on [Ergen 2004].

ZigBee supports three network topologies: the *star topology* and the *peer-to-peer topology* shown in Figure 5.4, and the *cluster-tree topology* shown in Figure 5.7.

A ZigBee network has two types of devices—a *full-function device* (FFD), and a *reduced-function device* (RFD). An FFD can function as a wireless router. It can communicate with other FFDs and RFDs. An RFD on the other hand is a simpler device with less functionality. It can communicate with an FFD but not another RFD.

The formation of an ad hoc ZigBee network starts with an FFD appointing itself the *PAN Coordinator* of a new ZigBee network. It also assumes the role of the head of the cluster—or *cluster-head*—and starts broadcasting a beacon periodically. Upon receiving the beacon, a nearby device that wishes to join the cluster seeks the permission of the PAN coordinator. If permitted the new device joins the cluster, adding the cluster-head as its parent. The cluster-head adds the new device as a child. As new devices are added to the growing cluster the FFD devices in the cluster, which can broadcast beacons, continue to recruit additional devices to the cluster.

A PAN coordinator is responsible for managing the entire network. It maintains awareness of whether a node in the network is an RFD or FFD node. The PAN coordinator is also responsible for communicating data from the network to agents outside the network. The memory, computation, and power resources needed in a PAN coordinator are therefore considerably higher than for other nodes in the ZigBee network. Figure 5.7 illustrates an example of an ad hoc ZigBee network, with the numbers on the nodes indicating the order in which the nodes joined the network.

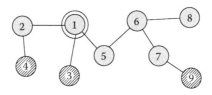

FIGURE 5.7
An illustration of the cluster-tree topology. The shaded nodes represent full-function devices, the striped nodes are reduced function devices. The specially marked node 1 is the PAN coordinator.

Mobile Web

Increasingly, the web is being accessed by mobile devices such as smart phones, laptops, and personal digital assistants. The mobile devices open new possibilities for web applications. For example, geo-location of a GPS-enabled mobile device can be used by search engines to sort the search results by geographic proximity to the current location of the mobile device.

While embedded hardware, such as GPS, opens new possibilities, the limitations of the mobile device pose serious challenges to the delivery of web content. For example, the small screen size and the low data transmission rate of mobile devices constrain the web content intended for them to be more focused, to avoid degradation in the end user's experience. Further the lack of a pointer, such as a mouse, in a mobile device imposes the constraint of designing web content without relying on the functionalities of a mouse. Also, mobile devices with small displays make it difficult, if at all possible, to multitask with separate windows. The small screen size also limits the amount of information that can be fit into a displayed page. For reasons such as the above web content intended for mobile devices involve different design criteria than those used for web content intended for standard desktops.

In response to the needs of mobile devices W3C launched the *Mobile Web Initiative (MWI)* to promote the development of web content suited for mobile devices. MWI compiles the best practices and guidelines for delivering web content to mobile devices. It also promotes a wider use of the developments in markup languages (e.g., HTML5) and style sheets (e.g., CSS3) for customizing web content to the target devices [W3C 2012].

There are two contrasting paradigms for adapting the web content to make it more friendly for mobile devices. One approach is to design *device-independent content* and have the server customize the contents that are delivered to the mobile devices through *content negotiation* with the device's browser. The other approach is to design *device-dependent content* that is tailored to mobile devices. A new top-level domain, *.mobi*, has been established to pursue the latter approach. The domain is managed by the *mTLD* registry, which runs a website called *mobiForge* (http://mobiforge.com/). *mobiForge*, an independent resource for developers who design content for mobile devices, also offers a free testing tool called *Ready.mobi* that website designers can use to check whether their websites are ready for mobile devices. Web sites registered under the .mobi domain are required to conform to the guidelines established by the mTLD registry [.mobi 2012].

Tim Berners-Lee, the creator of the World Wide Web, has criticized the emergence of device-dependent content. His criticism is supported by the W3C Technical Architecture Group and is best expressed in his own words

[Berners-Lee 2004]. An excerpt is quoted below, for it also provides an eloquent reiteration of the need for universality of the web.

> *"The Web must operate independently of the hardware, software or network used to access it, of the perceived quality or appropriateness of the information on it, and of the culture, and language, and physical capabilities of those who access it [WTW].* Hardware and network independence in particular have been crucial to the growth of the Web. In the past, network independence has been assured largely by the Internet architecture. The Internet connects all devices without regard to the type or size or band of device, nor with regard to the wireless or wired or optical infrastructure used. This is its great strength. From its inception, the Web built upon this architecture and introduced device independence at the user interface level. By separating the information content from its presentation (as is possible by mixing HTML with CSS, XML with XSL and CSS, etc.) the Web allows the same information to be viewed from computers with all sorts of screen sizes, color depths, and so on. Many of the original Web terminals were character-oriented, and now visually impaired users use text-oriented interfaces to the same information.*
>
> *... It is true that to optimize the use of any device, an awareness on the part of the server allows it to customize the content and the whole layout of a site. However, the domain name is perhaps the worst possible way of communicating information about the device. Devices vary in many ways, including:*
>
> - *Network bandwidth at the time,*
> - *Screen size and resolution,*
> - *Intermittent or continuous connectivity,*
>
> *and so on. While with the current technology, the phrase [sic] Mobile may equate roughly in many minds to something like a cell phone, it is naive—and pessimistic—to imagine that this one style of device will be the combination that will endure for any length of time. Just as concepts such as the Network PC and the Multimedia PC, which defined profiles of device capability, were swept away in the onrush of technology, so will an attempt to divide devices, users, and content into two groups.*
>
> *The Web works by reference. As an information space, it is defined by the relationship between a URI and what one gets on using that URI. The URI is passed around, written, spoken, buried in links, bookmarked, traded while Instant Messaging and through email. People look up URIs in all sorts of conditions.*
>
> *It is fundamentally useful to be able to quote the URI for some information and then look up that URI in an entirely different context. For example, I may want to look up a restaurant on my laptop, bookmark it, and then, when I only have my phone, check the bookmark to have a look at the evening menu. Or, my travel agent may send me a pointer to my itinerary for a business trip. I may view the itinerary from my office on a large screen and want to see the map, or I may view it at the airport from my phone when all I want is the gate number.*

---

* Berners-Lee, T. Weaving the Web, Harper, San Francisco, 1999.

> *Dividing the Web into information destined for different devices, or different classes of user, or different classes of information, breaks the Web in a fundamental way.*
> *I urge ICANN not to create the .mobi top-level domain."*

Summary

The mobile Internet paradigm emerged in response to the need of a growing number of mobile devices to remain wirelessly connected to the Internet while roaming. We have presented selected basic notions about the mobile Internet, mobile ad-hoc networks, and mobile web that are germane to our later discussion of I-2. The prototype for I-2, presented in Chapter 11, draws upon the discussion on wireless protocols and MANETs presented in this chapter.

6

Internet of Things

The vision of bridging the cyber and physical worlds to forge an integrated infrastructure is more than a decade old. Such an infrastructure, in which the "things"—everyday physical objects—are connected to the Internet was christened the *Internet of Things* (IoT). As we outlined in the preface, and discuss further in Chapter 10, connecting the "things" to the Internet is not the central issue in building the new infrastructure.

Connecting things to the Internet leads to an entropic explosion, owing to the overwhelming heterogeneity of objects and digital resources that IoT will encompass—from safety pins and coffee mugs to airplanes and bridges, from pictures and audio clips to national debt statistics and information on migratory patterns of monarch butterflies. The resulting cyber-physical system would comprise an awkward conglomeration of disparate and often functionally incompatible entities.

The critical issue is the development of a gluey framework, an ecosystem, in which the seemingly disparate physical and cyber resources are not only connected, but are functionally woven together into a giant infrastructure in which all the entities interoperate seamlessly. A successful integration of the cyber and physical worlds should result in a planetwide infrastructure in which the boundaries between the cyber world and the physical world gradually fade way as the infrastructure begins to function as one monolithic giant organism.

The envisioned planetwide cyber-physical system would be a network of end nodes that interact with each other. That some of these end nodes could encapsulate physical things while others encapsulate digital resources or even humans is of secondary importance. Hence, the envisioned infrastructure is renamed *Internet 2.0*, or *I-2*, in this book to underscore that the physical things are not the central entities within the infrastructure. However, a considerable amount of previous work that is relevant to our discussion has been done under the umbrella called the Internet of Things. We review selected aspects of such previous efforts in the following sections. In this chapter we have reverted to using the term Internet of Things in deference to the previous efforts.

Objective of Internet of Things

Kevin Ashton, one of the cofounders of the Auto-ID Center, was also one of the early proponents of the notion of Internet of Things. He envisioned an infrastructure in which computers would be empowered to obtain data about the physical world without human intervention. His perspective on IoT is best expressed in his own words, quoted below [Ashton 2009]:

> *"The fact that I was probably the first person to say Internet of Things doesn't give me any right to control how others use the phrase. But what I meant, and still mean, is this: Today computers—and, therefore, the Internet—are almost wholly dependent on human beings for information. Nearly all of the roughly 50 petabytes (a petabyte is 1,024 terabytes) of data available on the Internet were first captured and created by human beings—by typing, pressing a record button, taking a digital picture or scanning a bar code. Conventional diagrams of the Internet include servers and routers and so on, but they leave out the most numerous and important routers of all: people. The problem is, people have limited time, attention, and accuracy—all of which means they are not very good at capturing data about things in the real world.*
>
> *And that's a big deal. We're physical, and so is our environment. Our economy, society and survival aren't based on ideas or information—they're based on things. You can't eat bits, burn them to stay warm or put them in your gas tank. Ideas and information are important, but things matter much more. Yet today's information technology is so dependent on data originated by people that our computers know more about ideas than things.*
>
> *If we had computers that knew everything there was to know about things—using data they gathered without any help from us—we would be able to track and count everything, and greatly reduce waste, loss, and cost. We would know when things needed replacing, repairing, or recalling, and whether they were fresh or past their best.*
>
> *We need to empower computers with their own means of gathering information, so they can see, hear and smell the world for themselves, in all its random glory. RFID and sensor technology enable computers to observe, identify, and understand the world—without the limitations of human-entered data."*

Ashton viewed the IoT as the technology that empowers computers to *sense* the physical world without human intervention. The initial pioneering vision of Ashton's has evolved even further in the years that followed. While there is no universal consensus about the objectives of IoT, currently IoT is expected to have a larger functionality. It is expected to empower computers not only to *sense* the physical world but also *actuate* objects in the physical world, that is, do tasks without human intervention. Further, IoT is also expected to provide a platform that can *mediate* interactions among objects without human intervention. For example, IoT would enable a sensor that detects a gas leak in a residence to shut off the furnace and alert the gas company without waiting for human intervention. The actuation of physical objects, such as the furnace, and the interactions between objects, such as the interaction between gas sensor and the furnace, may not add to the stored data on the Internet and yet such interactions are encompassed in the current vision of the IoT.

IoT, described above, is a somewhat awkward infrastructure because it embodies a contrived distinction between physical and nonphysical worlds. The distinction between physical and nonphysical resources, though important from the implementation perspective, is not of importance either from a user's or from an architectural perspective.

For example, consider an online user who interrogates a retailer's web site about the availability of an item. There are two possible scenarios at the retailer's end. Either the requested information is available as a cyber resource—that is, as an entry in the retailer's inventory database. Or the information is not available in a database but rather is in the physical world—the stock of RFID-tagged items on the retailer's shelves—and the requested data is obtained by interrogating the tagged items using RFID readers. The user does not really care how the lookup service is implemented at the retailer's end. Whether the retailer consults a cyber resource—database—or physical resources—items on shelves—before returning the requested information is an implementation detail that is of interest to the retailer, but not the user. Such details must be hidden behind the lookup service provided by the retailer and should not be allowed to enter the discussion about the architecture that supports customer-retailer interactions.

In a generalization of IoT that we have called I-2, physical and nonphysical resources are treated on equal footing as components encapsulated inside a service agent. A service agent is taken to be an entity that provisions and/or consumes a service. The atoms—the building blocks—of I-2 are neither physical nor nonphysical resources but rather the dynamic interacting *service agents*, which could be implemented using physical and/or nonphysical resources. I-2 is discussed at length in later chapters. This chapter focuses on IoT, the technologies and activities that are converging to build IoT, the progress that has been made to date, the challenges that remain, and selected applications that are indicators of what lies ahead.

Overview of the State of the Art

At present the IoT does not exist as a sprawling seamless global infrastructure that it is envisioned to be. Rather it can be found as isolated networks of various sizes that are largely operating as disconnected islands of activity. Starting with the pioneering creation of the Auto-ID laboratories by

Sanjay Sarma, David Brock, and Kevin Ashton in 1999 [Sarma et al. 2000] and extending up to the recent commitment of about $800 million by the Chinese government [Yan 2011], significant investments have been made and are being made to build the IoT. However, the efforts to build IoT continue to be fragmented, and more than a decade later, we do not have even a serious prototype, let alone an operational global IoT infrastructure. The glacial progress signals that some systemic structural obstacles might be impeding the emergence of the paradigm. In Chapter 7 we take a closer look at the comparative histories of the Internet on the one hand and the IoT on the other in an effort to uncover the structural barriers that might be hindering the emergence of IoT.

The following paragraphs present an overview of the current status of IoT. The discussion is not comprehensive. It is meant to provide a glimpse of the various activities geared toward building the infrastructure. The discussion is organized by the scale of the efforts, ranging from the activities that involve hundreds of participating organizations down to the efforts of single organizations and laboratories.

EPCglobal Network

EPCglobal, a multiorganization collaboration operating under the umbrella of GS1, is engaged in developing an infrastructure called the *EPCglobal Network* to support interoperability of the supply chains of the participating commercial organizations [Balkesen 2008]. Objects in a supply chain are tagged with RFID transponders labeled with Electronic Product Codes (EPC). See Chapter 2 for a description of EPC. The EPC enables item-level tagging. That is, two different objects of the same object class and labeled by the same EPC manager would have EPCs that agree in all fields except the serial number field.

Upon reading the EPC of a tagged object, the information about the object is retrieved through a two-step process. In the first step, the EPC is used to determine the location of the database containing the information about the EPC. Such a database and its associated service layers is part of the so-called EPC Information Service (EPCIS) of the EPC Manager. The mapping from the EPC to the location of the corresponding database is achieved by interrogating a network of servers that comprise the Object Naming Service (ONS) arm of the infrastructure. The database, located using ONS, is then interrogated to obtain the necessary information about the tagged object. (See Figure 2.6.)

In addition to the information about the tagged objects—which is static data that do not change over the lifetime of the object—the EPCIS is also used to store dynamic data about the events involving the tagged objects. For example, as a tagged object flows through the supply chain from the manufacturer, through the various distributors to its final destination, readers located at intermediate checkpoints record the EPC of the object. The dynamic data recorded by the readers are stored in the EPCIS at the intermediate checkpoints. The EPC Discovery Service (EPCDS) is a service expected to be provided by the EPCglobal Network infrastructure to enable a trading partner to query the entire network and obtain a full history of all the events pertaining to an EPC. The EPCDS is expected to accord vendors a greater control over the supply chains by enabling efficient implementations of services such as product recalls. The EPCDS is also expected to help safeguard against counterfeiting [Lorenz et al. 2011]. The entire distributed infrastructure comprising the ONS, the EPCIS, EPCDS, and the associated data standards and communication protocols constitutes the *EPCglobal Network.*

The EPCglobal Network, which enables interaction between the physical objects in supply chains and the information infrastructures that are tasked to manage the supply chains, is an example of a large cyber-physical infrastructure. The architecture of the EPCglobal Network and its implementation have evolved for over a decade and have benefited from the efforts of the *EPCglobal* and its participating member organizations. EPCglobal Network, however, is designed to operate in a narrow domain. It is largely geared towards promoting interactions with supply chains and is not an open general-purpose infrastructure that an IoT is envisioned to be.

Ubiquitous ID Network

The EPC, described above, is largely slanted toward a single application— supply chain management. The fields of the EPC are formatted to contain information pertinent to supply chains, such as the identity of the manufacturer, the type of product, and the serial number of the product. The EPC is not readily adaptable to label general objects, such as a plant or a human or even web pages. The EPC cannot be used to label abstract entities such as concepts. On the other hand, the *uCode* (ubiquitous Code), supported by the Ubiquitous ID Center in Japan, is a general-purpose labeling scheme that can be used to label any entity of interest [uID 2012].

uCode is a 128-bit number that can be associated with any resource. It is tag-agnostic. That is, the uCode can be used with any tagging scheme including bar codes, two-dimensional QR codes, and RFID tags. The use of uCode is supported by an Internet-based infrastructure. Underlying the infrastructure is the uID architecture that is patterned after the Internet architecture [uID Center 2006]. The uCode technology and the uID architecture provide a viable addressing scheme for I-2. It has been deployed in the Tokyo Metropolitan Assembly Building in midtown Tokyo, art museums, and a zoo to provide users an immersive experience in their interactions with the physical world around them [uCode 2012].

Another domain in which a large number of industrial partners have converged to bridge the cyber and physical worlds and to promote global interoperability is the IPSO Alliance.

IPSO Alliance

Smart devices such as the smart thermostats and the smart energy meters, discussed in Chapter 2, are slowly percolating into everyday life. The different devices on the market, however, do not use a common communication protocol and are often based on proprietary protocols. The incompatibility of the communication protocols is a serious barrier to their interoperability. Absent a global organization to force convergence, the divergent business interests of the individual manufacturers will likely lead to a proliferation of incompatible protocols. The *IPSO (Internet Protocol for Smart Objects) Alliance* emerged as a multiorganization initiative to combat such divergence and promote interoperability among smart devices.

The IPSO Alliance proposes using the Internet Protocol (IP) as the common communication platform for all smart objects.[*] The alliance, comprising more than 60 member companies, seeks to advance its mission mainly by organizing global interoperability events called Interops. In Interops IP-based smart devices, spread across the globe and produced by different manufacturers, are shown to interact seamlessly with each other. For example, in the Interop event held during the IETF's 84th meeting [IETF 2012] IP-based devices from different vendors, spread across Canada, Finland, France, Sweden, and the United States, were demonstrated to work seamlessly. The Interops seek to demonstrate that IP provides a scalable and stable option that can serve as a substrate for a global infrastructure of smart devices. In addition to Interops the IPSO alliance also publishes white papers and case studies to spread the use of IP and outline the market opportunities for IP-based smart devices. It also works with standards organizations, most notably the Internet Engineering Task Force (IETF), to support the development of standards that impact the use of IP in smart objects.

While the IP is a pervasively used protocol, being resource-intensive it was considered unsuitable for resource-constrained smart objects. However, the

---

[*] Communication protocols, including the *Internet Protocol*, are discussed in Chapter 5.

emergence of 6LoWPAN, which implements IP with a lower resource requirement, has demonstrated the feasibility of using IP in resource-constrained
smart objects. For further details about the IPSO alliance the reader is
referred to IPSO [2012].

A similar initiative related to smart systems is the European technology
Platform for Smart Systems integration (EPoSS). Several member nations, as
well as industrial and academic partners in the European Union, have converged to establish EPoSS. The objective of EPoSS is to consolidate and streamline the activities related to the integration of smart systems [EPoSS 2013].

Monitoring the Earth and Its Atmosphere

The *Group on Earth Observations (GEO)* is a partnership that includes, as of
March 2012, 88 governments, the European Union, and 64 organizations.
Its mission is to pool the observations about the earth gathered from the
elaborate network of instruments and sensors owned and operated by the
partners—such as temperature sensors in buoys on oceans, seismic sensors,
and satellites that monitor the earth's environment—to facilitate recovery
from disasters and support global decision making that impacts the climate,
energy and water consumption, human health, agriculture, conservation of
biodiversity, and the management of various ecosystems [GEO 2012].

The GEO is building the *Global Earth Observation System of Systems (GEOSS)*
infrastructure to advance its mission. GEOSS provides a framework in which
the data derived from the instruments and sensors that are observing the
earth and its atmosphere are consolidated to provide decision-aiding information over the Internet. GEOSS is a sprawling, automated data acquisition
infrastructure feeding real-time information about the physical world into
the cyber infrastructure. Although it is not a general-purpose infrastructure,
GEOSS enhances the Internet in the spirit of I-2 by connecting the Internet
to the instruments and sensors, enabling the Internet to acquire information
about the physical world without human intervention [GEOSS 2013].

The *Central Nervous System for Earth (CeNSE)* initiative launched by Hewlett
Packard resembles the GEOSS initiative in that it also aims to observe the
physical environment of the earth. In the CeNSE initiative Hewlett Packard
hopes to deploy about a trillion sensors and actuators all over the earth. The
sensors are expected to report on aspects of the physical world, ranging from
the structural health of buildings and bridges to seismic activity. In a related
collaboration with Shell Oil Company, Hewlett Packard is applying its sensor (accelerator) technology to obtain a high-fidelity mapping of the hydro-
carbon reserves on the earth [Mullins 2010].

NASA and Cisco Systems Inc. are collaborating to develop the *Planetary Skin*, an infrastructure that uses the Internet, orbiting satellites, and land-based, sea-based, and airborne sensors to obtain and analyze the data about the earth and its environment. In a pilot project, called *Rainforest Skin*, the collaboration is attempting to gather and analyze data about the carbon dioxide levels in the major rainforests, such as those in the Amazon basin [Burnham 2009].

National Initiatives

The European Union (EU) has sought to promote the development of the IoT through funding channels such as its Seventh Framework Program (FP7) [FP7 2012]. The IoT-I project supported by FP7 is aimed at developing a cohesive vision for IoT within EU, and also prime the social and economic environments within EU for the emerging IoT [http://www.iot-i.eu/public]. The IoT-A project on the other hand is aimed at developing the architecture for IoT [http://www.iot-a.eu/public]. The European Research Cluster on the Internet of Things (IERC) is seeking to coordinate the various IoT-related research activities within EU [www.internet-of-things-research.eu]. The research projects and other ongoing activities are described at [http://www.internet-of-things.eu/introduction].

China's strategic interest in IoT and its investment are being driven by a top-down policy. Starting in 2009, IoT has been identified as a technology of strategic national interest, sparking a competition among the regional governments to build the so-called IoT Model Cities. The initiatives launched by regional governments are ushering the IoT technology into aspects of everyday life such as transportation, city management, and information security [Inoue et al. 2011]. In 2011, the Chinese government announced an investment of about $800 million into developing the IoT in China over a 5-year period [Yan 2011].

Japan has embarked on a three-phase approach to build the infrastructure and promote widespread adoption of the IoT technology. In the first two phases, named e-Japan Strategy I and II, spanning 2001–2005, Japan's Ministry of Internal Affairs and Communication embarked on improvement of its narrowband and broadband network infrastructure, laying the hardware foundation for the modernization of its Information and Communication Technology [MIC1 2012]. In the third phase, named *u-Japan Policy*, starting in 2005, the network infrastructure is being used to create a ubiquitous networked environment that connects the cyber and physical infrastructures [MIC2 2012].

Enabling Platforms

In the early days of the web there was a marked distinction between content producers and content consumers. Commercial and academic organizations created the web pages and populated them with content. The contents of the web pages were consumed by individual users of the web. Over the years the web has evolved into what has come to be called the *Web 2.0* in which the individual users both consume and generate the contents of the web. The magnitude of Web 2.0 is evident in the explosion of web content on social media websites and the number of web pages maintained by individual users of the web. *Crowdsourcing* the generation of content has accelerated the growth of the web.

Taking the cue from Web 2.0 some efforts have focused on accelerating the growth of IoT by crowdsourcing its evolution [DiYSE 2009, Pfister 2011]. Currently, one needs considerable technical expertise to participate in the task of building the infrastructure and applications for IoT. Consequently, a relatively small set of researchers and developers are currently engaged in the effort. If, on the other hand, user-friendly application development environments were made available, then even individual users who lack technical expertise would be enabled to participate in building the infrastructure. Several such application development environments are being built. A few representative examples are discussed below. The survey is not comprehensive. The examples were chosen to illustrate the spectrum of efforts that are underway.

Figure 6.1 shows how a wireless device such as a wireless-enabled thermostat can be configured to make it accessible over the Internet. The architecture shown in Figure 6.1 resembles the architecture used by the *Arrayent Internet-Connect Platform* [Arrayent 2012] and the *ioBridge* Platform [ioBridge 2012]. A gateway, capable of communicating wirelessly with the

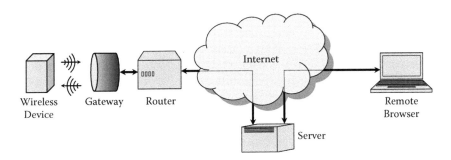

FIGURE 6.1
Architecture for web-enabling wireless devices.

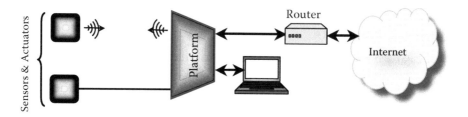

FIGURE 6.2
An interface platform for programmable control of sensors and actuators.

device, connects to the Internet through a router, thereby establishing an Internet-based communication between the gateway and a server on the cloud. The server acts as an intermediary that facilitates communication between a remote browser and the wireless device. Information about the device can be retrieved by and commands to the device transmitted from the remote browser through the server. Technologies such as the Arrayent platform provide a relatively simple plug-and-play capability to web-enable wireless devices.

Proceeding to a lower level, there are hardware platforms that interface a computer to the sensors and actuators, thereby giving users a greater degree of control to program the behavior of sensors and actuators. Figure 6.2 illustrates the architecture of such platforms. Two such platforms are *Arduino* [Arduino 2012] and *openPicus* [openPicus 2012].

The hardware platforms, such as Arduino and openPicus, contain a *microcontroller*[*] and provide an interface between tags, sensors, and actuators on the one hand and the computer on the other. The attractive feature of these interface platforms is that they abstract low-level hardware communications with the devices into constructs of a high-level programming language. *Arduino*, for example, provides a high-level Arduino Programming Language, using which the behavior of sensors and actuators connected to the platform can be programmed. The basic Arduino Uno microcontroller board can be connected to the Internet using an Arduino Ethernet Shield. The *openPicus* platform also offers the capability to interface with sensors and actuators. In addition, it also offers built-in connectivity to the Internet and an onboard web server as well. Thus, remote browsers can communicate with the openPicus's hardware module called Flyport directly without intermediary servers as in Arrayent or ioBridge.

Platforms such as *EVRYTHNG*, are seeking to build a "Facebook for Things" in which things, alongside humans, are endowed with identities on the web [Evrythng 2012]. The *Sen.se* initiative expands the notion of IoT to *Internet of Everything* in which humans, physical and cyber resources can interact seamlessly [Sen.se 2012]. The *Open.Sen.se* platform is envisioned to

---

[*] A microcontroller is a small computer that has a processor, memory, input/output ports, and serial lines for bidirectional data exchange with a microprocessor.

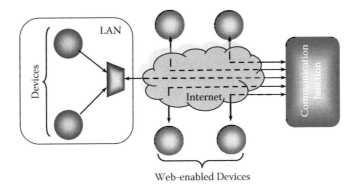

FIGURE 6.3
Communication junction to facilitate interaction among devices.

enable people to realize the above vision. The *ThingWorx* platform is also designed to enable people, systems, and physical objects to interact on equal footing in a seamless ecosystem [ThingWorx 2012]. One of the challenges in building a platform that can support the interaction of heterogeneous devices is providing the devices a means to communicate with each other. One option is to have a communication junction as shown in Figure 6.3. If a device needs to send a message to another device, the message is routed to the communication junction, which in turn routes the message to the target device. Such a junction can serve both web-enabled devices and local networks. The *iDigi* platform provides such a service to support interaction among devices [iDigi 2012].

The above examples are representative of the platforms that are enabling individual users to participate in building I-2. Other platforms of interest include *Nimbits, ThingSpeak, Exosite,* and *Manybots* [Doukas 2012].

Selected Applications

We present selected applications that showcase the efforts to integrate the cyber and physical worlds toward building IoT. The applications span multiple scales—ranging from personal intranets of things to massive infrastructures such as supply chains of global retailers. In the following discussion, the applications are organized by scale. The particular products that are described below could have several competitors on the market. The examples presented below do not constitute a comprehensive survey but are intended to give the reader a flavor of IoT applications. The material in this section, by its very nature, is expected to become obsolete. However, the ideas the applications embody are expected to have enduring value.

Corventis [Corventis 2012] has developed a light, wireless, water-resistant, adhesive sensor, called *PiiX Monitor* that can be stuck to a patient's chest. The sensor continuously monitors cardiac activity, respiration activity, body fluids, body activity, and posture (using a built-in accelerometer). The physiological data obtained by the sensor is then transmitted from the PiiX Monitor to a wireless transmitter device, called *zLink*, placed in the vicinity of the sensor. BlueTooth communication is used for data transmission from PiiX to zLink. zLink in turn routes the data to Corventis servers using cellular communication. Certified technicians at the Corventis Monitoring Center monitor the sensor data initiating follow-up actions if predefined trigger events are detected. The application is an example of how the wireless sensing technology is being used to continuously transmit data about an object—the human body—to the cyber infrastructure.

While the PiiX Monitor extracts data for diagnostic purposes, another product called *GlowCaps* tracks patients' adherence to medication. GlowCap is a replacement for the standard cap used for pill bottles. Unlike a standard cap, however, GlowCap houses a small computer that tracks when the bottle is opened. The bottle usage data is then routed through a local gateway device, deployed inside the residence, to a secure network and stored in a private data warehouse. The aggregate statistics about medication adherence are reported every month to the patient. The application is an example of how wireless sensing is being used to extract data, not only about physical objects, but about the processes, such as medication adherence, occurring in the physical world.

Like GlowCap, the *Plogg,* is a device intended for use in home or office [EOL 2012]. It is a digitally enhanced replacement for a standard electrical plug. The *Plogg* is designed to track the electricity usage of the device plugged into it. The usage data is then transmitted wirelessly by the Plogg to either a gateway device or a local computer for data logging. The device is an example of how the sensing technology is being used to obtain information, not only about objects and processes but also entities such as energy.

The PiiX Monitor, GlowCap, and Plogg are examples of devices that offer nonversatile coupling to the physical world. The *Mirror RFID reader* on the other hand provides a general-purpose bridge between the physical and the cyber worlds. The reader plugs into a computer's USB port and can sense RFID tags brought into its vicinity. The computer can then be configured to launch an assigned application when a tag is detected in the vicinity by the reader. Thus, for example, when a child returns home from school the reader could be set up to detect the kid's tagged bag and send an email to parents. The device is an example of percolation of the RFID technology into households.

The Mirror reader is not the only technology gaining popularity in households. Home automation is a rapidly growing industry and several applications of IoT are targeting the home automation market. We mention a few examples below.

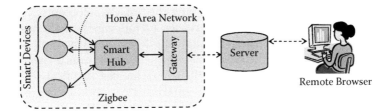

FIGURE 6.4
The architecture of a Home Area Network that enables remote monitoring of appliances, home security, and energy usage.

An interesting application is the *Botanicalls* service [Botanicalls 2012]. The Botanicalls system comprises a sensor that can detect the moisture content in soil. When the moisture content in the soil of say a potted plant, falls below a threshold the system triggers a phone call to the owner with an alert that the plant needs watering. The technology can be easily adapted to sense the moisture content in farm soil, giving farmers a valuable sensing capability. The architecture of the technology can be adapted for other environments, such as sensing flooding in basements. More recent versions of the Botanicalls system also offer the option of sending a text message or a Twitter message instead of placing a phone call.

Whereas the Botanicalls system is specialized to monitor plants, the *AlertMe* platform supports a range of devices that enable one to monitor the energy usage and security of a home from anywhere [AlertMe 2012]. The architecture of the AlertMe system is shown in Figure 6.4.

The AlertMe system deploys a Home Area Network. At the core of the network is a SmartHub that interacts with and manages the smart devices in the network, using Zigbee technology. AlertMe offers the following smart devices: (1) *SmartEnergy Meter* that clamps to the electrical wiring near the standard electricity meter and gathers the electricity usage data. The usage data is then transmitted to both the SmartHub and the company that supplies electricity; (2) *SmartMotion sensor* that can be deployed anywhere in the house. It monitors the home for movements sending data about unusual movements to the SmartHub. The SmartHub can be programmed to sound an alarm and send either a text message or an email, or place a phone call when suspicious movements are detected; (3) *SmartContact sensor* that can be attached to doors and windows, which if opened will cause the sensor to trigger an alarm and a text message, an email, or a phone call to the owner; (4) *SmartAlarm sensor,* which listens to the alarms in the house, such as that from a smoke detector, or carbon monoxide detector, alerting the owner about the alarm through an email, a text message, or a phone call; (5) *SmartPlugs,* which replace the standard plugs, and can be turned on or off over the web. The SmartPlugs, like Plogg, can also record the energy usage of the appliance plugged into it; (6) *SmartKeyfob,* whose presence can be detected by the SmartHub. When a family member, such as a child, carrying the SmartKeyfob enters the house the SmartHub can detect the entry and send a message over the web.

The SmartHub is connected through the home gateway to the AlertMe server, which resides on the cloud. The connection permits a two-way communication between the server and the smart devices. Applications running on the server can pull data from the smart devices and also transmit commands to devices such as the SmartPlugs. The owner in turn can receive messages from, monitor, and control the smart devices from anywhere in the world by connecting to the server either through a computer or a smart phone. The AlertMe system illustrates how the physical surroundings within a home can be sensed and controlled using wireless bridge technologies.

While the AlertMe system monitors energy usage in a home the popular *Nest* thermostat works to actually reduce the energy usage while enhancing the living comfort [Nest 2012]. The Nest thermostat is distinguished by its ability to learn about a user's preferences over time and adapt itself to deliver the user's preferred temperature settings at different times of the day. It comes bundled with several advanced features geared towards minimizing the total energy usage. For example, the built-in activity sensors detect if the home is vacant. The thermostat is connected to the web, allowing it to pull the weather forecast data. Forecasts about the external temperatures are translated into control decisions that minimize energy usage. Finally, it can be accessed over the web giving a user the flexibility to monitor and program the thermostat from a remote location. The capability to control the thermostat over the web enhances the living comfort, especially in extreme weather conditions.

The applications discussed above operate largely in a residential or office setting. Next, we consider applications on a slightly larger scale. A notable citywide application involves tracking the locations of public buses, in real time, as they ferry passengers through the city. GPS trackers installed on buses can be used to periodically determine their locations. The location data is then transmitted wirelessly to the servers in the central office. By accessing the real-time data feed applications running on smart phones can accurately predict the arrival times of buses at various stops [MBTA 2012]. The *CatchTheBusApp* provides such a service for Boston's MBTA bus system as well as San Francisco's Muni system [CatchTheBussApp 2012]. It works on most smart phones, providing users an accurate prediction of the arrival times of the buses. This application shows how the ability to obtain real-time information about physical objects, such as buses, translates to enhanced experience for the passengers.

The previous application exploits real-time location data about the moving buses. Even real-time information about stationary vehicles can be used to provide a valuable service, as Streetline's *ParkSight* application illustrates [Streetline 2012]. Streetline's sensors, installed at the road-side parking spots in a city are designed to sense the arrival/departure of a vehicle at the parking spot [Streetline 2012]. When the parking spot is occupied or vacated, the sensor transmits the information about the status change through a network of repeaters to a gateway, which in turn transmits the data to Streetline's servers. The real-time data about every parking spot in the city is made available to the ParkSight application that runs on smart phones. ParkSight uses the real-time data to display the locations of the vacant parking spots. The parking rules and fees at each spot are also made available to help the driver determine the parking spot of his/her choice. Streetline also offers similar services for parking garages, universities and airport parking facilities [Streetline 2012]. This application illustrates the enormous savings that can be realized by gathering the real-time data about the physical world. The following statistics [White Paper 2012, Shoup 2005] outline the magnitude of the potential savings.

It is estimated that between 8% and 74% of traffic congestion in downtown areas is caused by vehicles looking for parking spots. For example, 28% of the traffic in Manhattan and about 45% of the traffic in Brooklyn, New York, comprised drivers looking for parking spots. Studies spanning a 15-block area in Los Angeles (resp. Manhattan) revealed that drivers in the area travel about 950,000 (resp. 366,000) extra miles per year, pumping 730 tons (resp. 325 tons) of carbon dioxide into the atmosphere as a result.

Unlike the citywide applications such as CatchTheBusApp and ParkSight, the *Smart Grid* is a national infrastructure that makes the electrical distribution network in the physical world visible to the cyber infrastructure. The grid is the infrastructure (comprising the generators and the network of transmission lines and transformers) that delivers electricity from the power plant to the end users. The grid was designed to support a one-way flow of electrical power from the generators to the end users. The smart grid is an enhancement of the old grid in which, in addition to the one-way flow of power, the suppliers and the consumers are linked through a two-way communication channel. Old energy meters could record energy consumption but could not communicate the energy usage data. The new smart energy meters are empowered to record energy usage and communicate the usage data to both the consumer and the supplier in real time. The real-time usage data enables the supplier to layer the pricing structure to encourage consumers to shift their energy usage to off-peak periods. The information infrastructure operating alongside the distribution network also enables the utility companies to rapidly pinpoint the source of power outages. The ability to isolate the source of power disturbance enables the rerouting of power through the network so that end users who are not in the immediate neighborhood of the disruption are not affected by the disturbance. Thus, the communication infrastructure enables the smart grid to self-heal when parts of the network malfunction [SmartGrid 2012]. The earliest, and currently the largest, deployment of smart grid technology is in Italy. The infrastructure was built with an investment of 2.1 billion euros and is yielding annual savings of about 500 million euros, besides improving the quality of service

[NETL 2007]. The smart grid illustrates the magnitude of savings that can be derived by using IoT technology in the energy sector.

Standards for the Internet of Things

A more recent initiative aimed at catalyzing the development of IoT is the *IoT Global Standards Initiative* (IoT-GSI) launched by the International Telecommunication Union, a specialized agency of the United Nations focused on information and communication technologies. One of the objectives of the initiative is to develop the standards for IoT, consolidating the previous work done on standards development by other agencies [IoT-GSI 2011].

Summary

More than a decade after the birth of the notion of IoT, the paradigm exists today not as the global infrastructure that it was conceived to be, but rather as largely disconnected islands of intranets of things. These intranets span multiple scales, ranging from a personal intranet of things belonging to a single user to national infrastructures, like smart grids, and the trans-continental infrastructures such as the global supply chains. The various intranets that are operative remain mostly incompatible with each other. In Chapter 7 we compare the evolution of IoT with that of the Internet in an effort to identify the barriers that IoT may be facing.

Section IV

Internet 2.0 (I-2)

Overview

The National Intelligence Council identified the Internet of Things (IoT) as one of the six technologies that could have a disruptive impact on the United States as far ahead as 2025 [NIC 2008]. The European Union (EU) has been investing into IoT research through channels such as its Seventh Framework Programme (FP7). EU has also constituted a *European Research Cluster on the Internet of Things* to coordinate the EU-funded IoT-related efforts across Europe [IERC 2012]. Recently, China announced an investment of about $800 million to develop the IoT infrastructure within China [Yan 2011]. Japan's Ministry of Internal Affairs and Communication has adopted a u-Japan policy to build a network that integrates the physical and cyber infrastructures [MIC2 2012]. Seven Auto-ID laboratories, located in the United States, Asia, Australia, and Europe, are working with EPCglobal to architect the IoT [Auto-ID 2013]. A parallel effort is also being pursued by the Ubiquitous ID Center, based in Japan [uID 2012]. Besides these prominent activities, several other initiatives are also under way to facilitate the birthing of the IoT infrastructure.

IoT-related efforts have yielded several niche networks over the last decade. For example, the EPCglobal network has improved visibility in supply chains [Schuster et al. 2007], while the uID technology has been deployed in art museums in Japan to enhance visitors' interactions with the exhibits [uID 2013]. RFID-based electronic toll booths have expedited toll collection

[Banks et al. 2007]. These examples, and the others discussed in the previous chapter, however, represent fragmented and largely disconnected islands of activity. I-2 is yet to emerge as the general-purpose global infrastructure that it was, and is, envisioned to be.

The glacial progress on I-2 is puzzling. The hardware technology needed to build I-2 already exists. Pockets of activity focused on I-2 have mushroomed all over the globe over the last decade, bringing significant amounts of efforts to bear on the task of building I-2. I-2's progress can be calibrated by recalling that a decade after their inception the Internet and the World Wide Web had become sprawling operational networks. If the evolution of I-2 had mirrored the rate of the web's evolution, we would have a global I-2 infrastructure with an accreting user base by now. On the other hand, to date, we do not have even an operational prototype of I-2. Viewed against the backdrop of the evolution of the Internet and the web, it appears that the vision of I-2 is floundering.

To fathom why I-2 is struggling to gain foothold we turn to a related success story—the Internet—for clues. In Chapter 7, we take a closer look at the evolutions of the Internet and I-2. Internet's history is a case study in ingenious architectural design and skillful stewardship—all the way from the incubation phase, through the growth phase and finally into the commercialization and globalization phase. The birth of the Internet was midwifed by a cohesive group of visionaries, largely within academic research environment. The evolution of I-2, on the other hand, has deviated markedly from the course along which the evolution of Internet progressed. The design of I-2's architecture continues to be a work in progress without any clear consensus and marked by fragmented stewardship. If I-2 were the proverbial "broth," then far too many "cooks" are involved with the result that the vision has been splintered into divergent disconnected activities, without critical mass accreting anywhere.

In Chapter 8, we turn to the Internet and the web to identify the design principles that should be incorporated into I-2, if it is to emerge as a successful global infrastructure. The architectures of the web and the Internet share one common design feature that appears to be critical to their global success. Their architectures achieve dramatic reduction in complexity by using irreducible constructs—the *IP datagram* in the case of the Internet and the notion of *resource* in the case of the web—as the units of dynamic interactions among the end nodes. The simplicity at the core appears to be the key to making a global architecture robust and scalable, leading us to the question: can the architecture of I-2 also be rendered simple by using an irreducible construct as the unit of interaction between its end nodes? The required irreducible construct for I-2 is provided by the notion of the so-called ***web-enabled service***. Previous investigators have recognized the importance of services and service-oriented architectures to I-2 [Guinard, Ion, and Mayer 2011; Guinard et al. 2010; Rellermeyer et al. 2008]. However, we go a step further. We elevate *service* from being just a useful paradigm, and anoint *web-enabled service* as *the* irreducible unit of interaction between the end nodes of I-2. Taking the *web-enabled service* to be the irreducible unit of transaction and I-2 to be a network of *service agents* that provision and consume web-enabled services appears to yield the necessary simplification at the core of the I-2 architecture.

In Chapter 9, we discuss the notions of service and service-oriented architecture. We also discuss a relevant family of services, namely, *web services*. The architectural imperatives for I-2—the features that must be incorporated into I-2's architecture—are discussed in Chapter 10. Designing the architecture, however, is only a first step toward building I-2. The next steps are (1) translation of the architecture into an operational prototype, (2) deployment of the prototype across several noncommercial research communities, and finally (3) conduction of field tests on the prototype to iron out the wrinkles in the design. Deploying a prototype across several research communities can be prohibitively expensive if the resources needed for the prototype are to be acquired afresh. Faced with a similar challenge during Internet's incubation, the architects of the Internet exploited the existing resources such as the telephone networks and the satellite networks, using the existing networks as the backbone of the fledgling Internet, thereby reducing the investments required to incubate the Internet. In Chapter 11, we discuss the possibility of similarly building a prototype I-2 infrastructure using the available resources to reduce costs in the incubation phase. Finally, in Chapter 12, we discuss a roadmap for galvanizing progress on I-2. The discussion in the following chapters is targeted at a broad audience, including the policy makers, whose active involvement is critical for I-2's success.

7

Evolution of Global Infrastructures

The Internet, arguably one of the most successful global infrastructures, began its life humbly enough as just two ARPANET nodes—one at Leonard Kleinrock's Network Measurement Center at UCLA[*] and the other at Doug Engelbart's group at SRI.[†] In late 1969, the first host-to-host message was sent from UCLA to SRI heralding the birth of the infrastructure that would help usher the human race into information age [Leiner et al. 1997]. From a handful of users in late 1969, the usage of the Internet within the United States grew nearly a million-fold over two decades to about 2 million users[‡] in 1990 [World Bank 2011c]. With the advent of the World Wide Web around 1990, the Internet usage has seen a staggering rise. Figure 7.1 shows the growth in Internet usage within the United States since 1990. The worldwide statistics of Internet usage are also impressive. As of December 31, 2011, the world had about 7 billion people,[§] and about 2.2 billion Internet users[¶] [Internet Stats 2011]. More than three out of every four people in the United States and two out of every seven people in the world use the Internet today. The growth of the World Wide Web is just as impressive. In 2008, it was estimated that there were about a trillion different resources (URLs) on the web, or more than 140 resources for every person on the planet [Alpert and Hajaj 2008].

As impressive as the rate of growth of the Internet is, it is dwarfed by the rate at which another global infrastructure—*Facebook*—has grown over the last decade. Figure 7.2 charts the growth of Facebook from its inception in 2004 to October 2012, when the number of "monthly active users" on Facebook topped a billion.[**] In less than a decade, Facebook has attracted nearly one out of every seven people on the planet to its user base.

Another global infrastructure that has grown steeply, impacting aspects of modern world from entertainment to politics, is *YouTube*. Since February 2005, when its domain name *www.youtube.com* was registered, it has grown to become a cultural phenomenon. It was estimated that in 2011 YouTube

---

[*] University of California, Los Angeles.
[†] Stanford Research International, Menlo Park, California.
[‡] Estimate: 1,988,024 users.
[§] Estimate: 6,930,055,154 people.
[¶] Estimate: 2,267,233,742 users.
[**] I thank Stacy Cowley, Tech Editor at CNNMoney.com, for providing the data presented in this graph. The data were mined by her from Facebook's SEC Filings and its published timeline, which can be found at http://www.sec.gov/Archives/edgar/data/1326801/000119312512325997/d371464d10q.htm and http://newsroom.fb.com/content/default.aspx?NewsAreaId=20.

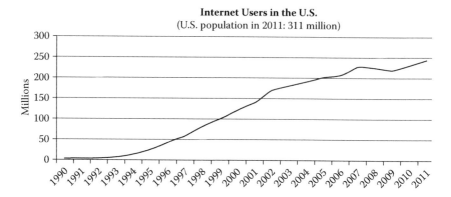

FIGURE 7.1
Number of Internet users in the United States. (From http://data.worldbank.org.)

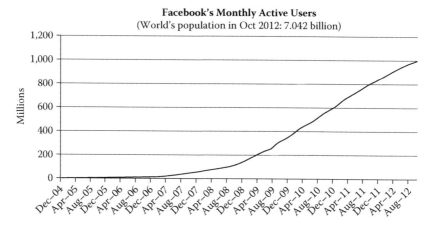

FIGURE 7.2
The growth of Facebook since its inception.

received about a trillion views [YouTube 2012a], which amounts to every person on the planet visiting YouTube once in three days.

The World Wide Web, Facebook, and YouTube are software infrastructures that operate over the Internet. Therefore, in comparing their growth rates with that of the infrastructures such as the Internet, which involve deployment of hardware, one has to correct for the disparity in the efforts, time requirements, and financial investments needed to grow the software and hardware infrastructures. But even factoring in such disparity, the above global infrastructures are spectacularly successful and have transformed the modern world.

In contrast, the I-2 has not followed a growth curve comparable to that of the above infrastructures. The vision of an integrated cyber-physical infrastructure in which the cyber infrastructure interacts with the physical world is at least a decade old. As the discussion in Chapter 6 shows, starting with the Auto-ID Center at MIT several commercial organizations and nations such as Europe, China, and Japan have converged to facilitate the development of the infrastructure. The technology needed to build the infrastructure is available. And yet, nearly a decade later, even an embryonic version of a global I-2 infrastructure does not appear to exist. As discussed in Chapter 6, there are fragmented islands of Intranets of Things that are largely disconnected from each other. Several of the intranets focus on the narrow range of needs that they were built to address. But none of the intranets appears to have the critical mass or momentum characteristic of an emerging global general-purpose I-2 infrastructure. Robert Williams, the editor of some of the ISO standards for RFID, articulates the disconnect between the evangelistic predictions about the Internet of Things and the ground reality in the following words [Williams 2008]:

> *Just under ten years ago … it was predicted that RFID tagged items and the "Internet of Things" would be ubiquitous by 2005. Yet here we are approaching 2009, and still waiting for something to happen. Why are we still waiting, and what is the real business case behind the "Internet of Things"?*

Why does the I-2 vision appear to be floundering? One of the suggested arguments for the rather slow diffusion of I-2 into everyday life is that the cost of RFID tag is still too high to permit its widespread use. Currently (late 2012), the cost of a passive tag has fallen to about 7 cents [RFID 2012], which is close to the 5-cents-a-tag target that has been pursued for many years [Ashton 2011]. And yet, I-2 has not gained momentum as one would have expected. So it seems that while low tag costs will certainly help, the tag costs were not the only hindrance to the birth of I-2.

Some argue that the business case for I-2 is weak even if the tag costs fall to 5 cents [Williams 2008]. The criticism about weak business case may be applicable to supply chains, where the cost of the tags impacts the profit margin. However, I-2 is not entirely about supply chains. It is also about an unprecedented integration of the world of everyday physical objects with the cyber infrastructure. It is a new paradigm, a new framework, much like the Internet was at its inception. There were very few returns on investment and hence there was a "weak business case" for the fledgling Internet as well during its incubation. It is hard to argue that the commercial potential of the Internet was evident during its early years, even before the birth of the World Wide Web. Full-fledged commercialization of the Internet did not start for more than 2 decades after its birth. However, the absence of returns on investment in the short term and a weak business case did not hinder the rapid growth of the Internet. And today, the business case for the Internet or the pivotal role it plays in the modern economy hardly needs discussion. Therefore, we have to look beyond the arguments about the business case to identify the reasons for the sluggish progress on I-2.

Just as the Internet sought to fold seemingly incompatible networks into a seamless giant interconnected infrastructure, I-2 seeks to fold heterogeneous entities—the cyber resources, physical resources, and humans—into a seamless networked infrastructure. Building I-2 involves adding hardware resources to the existing Internet to enable it to interact with the physical world. In that sense I-2 bears a greater similarity to a hardware infrastructure such as the Internet than it does to software infrastructures such as the World Wide Web, Facebook, or YouTube. Hence, the earlier question about the floundering vision of I-2 can be recast as two equivalent questions: Why is I-2 not following the Internet's growth curve? Why is I-2 floundering? To look for answers to the above questions, we start by recapitulating the history of the Internet.

Evolution of the Internet

The evolution of the Internet can be divided into four phases, arranged in chronological order as shown in Figure 7.3. The material for the following discussion is taken from Leiner et al. [1997].

Phase I: The first phase involved building a *prototype infrastructure* that would serve as a test bed for evaluating design alternatives. The construction of the infrastructure began with the development of the ARPANET, under the stewardship of Lawrence Roberts. The ARPANET was based on Leonard Kleinrock's new paradigm of *packet switching,* which continues to be at the heart of today's Internet. One of the bedrock principles of today's Internet is Robert Kahn's *open-architecture networking*—the philosophy that *"Each distinct network would have to stand on its own and no internal changes could be required to any such network to connect it to the Internet."* The foundations for the open-architecture networking were already being laid in the ARPANET, which used *interface message processors* to connect

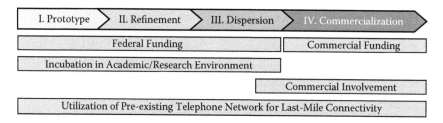

FIGURE 7.3
Early history of the Internet.

local networks of computers to the ARPANET, thereby separating the design of local networks from the architecture of the ARPANET. The interface message processors were designed under the stewardship of Frank Heart at BBN. The first node of the Internet—the first interface message processor—was installed at Kleinrock's Network Measurement Center at UCLA and the second at Doug Engelbart's group at SRI. The Network Working Group, stewarded by Stephen Crocker, developed the host-to-host *Network Control Protocol.* With the deployment of the Network Control Protocol, the prototype infrastructure was ready for field-testing. ARPANET, a network-of-networks, provided the architectural test bed for the development of the technology for interconnecting different networks, or the *Internet,* for short. The development of the prototype infrastructure lasted about six years, starting in 1966, when the plan for ARPANET was crafted, to 1972, when the implementation of the Network Control Protocol was completed.

Phase II: The second phase in the evolution of the Internet focused on *refinement of the prototype infrastructure* to make it robust and scalable. The Network Control Protocol, which depended heavily on the reliability of the underlying ARPANET and hence lacked host-to-host error control feature, was replaced with a more universal *TCP/IP* designed by Robert Kahn and Vint Cerf. The design of TCP/IP was also motivated by Robert Kahn's *open-architecture networking* philosophy, which sought to ensure interoperability among disparate local networks without restricting the architecture of the local networks. David Clark extended the early implementations of TCP/IP, which were for large time-shared systems, and developed TCP/IP for smaller desktop systems. As the size of the infrastructure grew, in tandem with the evolution of the LAN technology and the personal computer market in the 1980s, the prototype architecture was further refined to ensure scalability. Earlier, the names and addresses of all the hosts in the ARPANET were maintained in a single table. With the growing size of the infrastructure a distributed resolver service, the *Domain Name Service,* was developed by Paul Mockapetris. When the infrastructure had a few nodes all the routers were running the same routing algorithm. As the size grew, running the same algorithm on all the routers was no longer scalable, necessitating a *hierarchical routing model,* which laid the foundation for the *Interior* and *Exterior Border Gateway Protocols.*

Phase III: The third phase in the evolution of the Internet comprised *dispersion* of the TCP/IP protocol. TCP/IP was incorporated into *UNIX BSD* (Berkeley Software Distribution) operating system and disseminated widely within the academic community, which was the main user base of the emerging Internet. ARPANET mandated its users to migrate *en masse* from the Network Control Protocol to the TCP/IP by *January 1, 1983*. The TCP/IP was also the mandated protocol for the *NSFNET* that emerged in 1985. The dissemination of the TCP/IP in the software vendor community was orchestrated by organizing the *Interop Trade Show* in which selected vendors were given a forum to present the interoperability of their TCP/IP-based products with that of the others.

Phase IV: The last phase in the evolution of the Internet into a global infrastructure was the *commercialization* of the technology, stewarded by NSF. NSF partnered with DARPA to make the NSFNET and the ARPANET interoperable. The federal agencies also created the Federal Network Council to coordinate the support for the growing Internet infrastructure. The commercialization of the Internet was driven by two strategically important decisions made by NSF. In the first important decision, NSF encouraged the local networks on the NSFNET, which comprised mostly academic ecosystems, to open up the infrastructure for local use by the commercial customers. This enabled the nonacademic users to appreciate the commercial potential of the information infrastructure. At the same time, NSF's *Acceptable Use Policy* prohibited the use of the nationwide NSFNET infrastructure for commercial purpose. In the second important decision, NSF defunded the NSFNET in April 1995, forcing commercial organizations to take over the responsibility of enhancing and maintaining the global network infrastructure. The defunding of NSFNET forced the critical transition from a federally funded Internet to a commercially funded Internet.

The evolution of the Internet has progressed beyond the commercialization phase as discussed in the previous chapters. However, the later phases are less relevant to our objective, which is to identify the barriers to the birth of I-2. Therefore, we have restricted our attention to the evolution of the Internet from its birth to its commercialization. The above discussion is condensed for easy reference in Figure 7.3.

The salient features of the Internet's history are:

1. The efforts to create the new infrastructure, the Internet, began with the construction of a concrete *open prototype*. Field experience with the prototype served to expose the shortcomings of the original design necessitating architectural enhancements such as TCP/IP and DNS.

2. The prototype was conceived, built, and refined largely in a *research environment* and with *federal funding*. In the words of the people who architected the Internet [Leiner et al. 1997],

"The Internet is as much a collection of communities as a collection of technologies … This community spirit has a long history beginning with the early ARPANET. The early ARPANET researchers worked as a close-knit community to accomplish the initial demonstrations of packet switching technology …"

Since the incubation efforts were supported by federal funds the architects of the Internet were not saddled with the stymieing requirement of having to state the *business case* for the Internet even before it was designed and built. Not having competing commercial interests, they were not working at cross-purposes with each other. The commercial organizations were invited to expand the infrastructure only after its architecture was designed, tested, and was fully operational and the business case for it had become nearly self-evident.

3. The federal agencies, notably *DARPA* and *NSF,* made significant investments into building and promoting the prototype infrastructure. The visionaries who architected the Internet straddled research institutions and the funding agencies, fostering the research even as they provided strategic leadership to drive progress on the Internet. Lawrence Roberts, who stewarded the ARPANET at DARPA was also responsible for demonstrating the feasibility of wide-area networking. Robert Kahn, who was involved in the design of the interface message processors at BBN, and was one of the architects of TCP/IP, also served as the director of the Information Processing Techniques Office at DARPA. Again, Vint Cerf, who along with Kahn developed TCP/IP while at Stanford University, also served in DARPA. The cross-fertilization between the research community and the funding agencies is another noteworthy detail in the history of the Internet.

4. The *open-architecture networking* paradigm espoused by Robert Kahn has played a key role in the widespread adoption of the Internet. The architecture of the Internet was designed to be interoperable with any local network and placed no constraint on either the software or the hardware environments of the local networks that sought to connect to the Internet. The Internet was thus designed to be a *noncoercive meta-network* whose architecture was independent of the details of the local networks it interconnected.

5. The Internet was designed to have a *simple core architecture,* with all of the complexity and heterogeneity pushed to the edge. The Internet functions as a simple data transportation infrastructure, which does not care about the semantics of the data it transports, or the applications at the edge that generate and consume the data it transports. That is, the Internet was designed to be application-agnostic. As Vint Cerf remarks [Cerf 2010], the other networks such as the telephone

network and the television network were *purpose-designed*—that is, dedicated for a particular application. The Internet, on the other hand, was not designed for any application. Being a *general-purpose infrastructure* enabled the Internet to thrive and support ever more demanding applications over the years.

6. The data transmission on the Internet was designed to be on a *best effort* basis with no guarantee of success. Thus, the quality of service was *not* guaranteed in the Internet. The earlier communication technologies such as the circuit switching method, which sought to guarantee a quality of service, were rejected in favor of the packet switching method, which does not guarantee success in transmission. Paradoxically, the robustness of the Internet, its efficiency, speed, and scalability appear to stem from the decision to abandon guarantees about the quality of service.

7. *Security* was a secondary concern in the initial design of the Internet. In Vint Cerf's words [Cerf 2010],

"Although, in fact, we didn't focus very heavily at all on security. And we might look back on that and regret it, but I have to say that if we had tried to focus heavily on security in these early days we might never have even gotten anything built that we could test."

Although the initial basic version of the Internet, deployed mostly in academic environment, did not implement security features the importance of making the Internet secure was well recognized and efforts to build a secure version of the Internet had begun even in the early years [Vint Cerf, private communication].

8. The academic community provided the early user base for testing the Internet. In the early days the community's interest in the Internet was sustained to no small extent by the killer application—the *email*—that was devised by Ray Tomlinson of BBN. The email application knit the early users of the Internet into an engaged community that served as ready customers for the later enhancements of the infrastructure such as the World Wide Web.

9. The first wide-area network communication occurred between two computers located on either side of the United States over a *telephone line*. Telephone lines also served as the *backbone* of the Internet in its early days. Although the Internet backbone today has dedicated fiber optic lines that operate at significantly higher bandwidths, the last mile of communication from the ISP to subscribers, such as the individual home users, still relies on pre-existing infrastructures such as telephone and cable lines. Exploiting the existing networks, to which the customers are already connected, lowers the cost of

connecting to the Internet and facilitates a widespread adoption of the technology.

10. Finally, as Figure 7.1 shows, the *World Wide Web,* born in the period 1989–91, has contributed significantly to the growth in Internet usage. The web made the Internet more accessible to lay users. The users no longer had to contend with low-level communication details but could instead focus on exploiting the connectivity provided by the Internet to build a rich ecosystem of applications on top of the Internet. The web has played a critical role in Internet's transition from being an academic infrastructure to a commercial infrastructure.

Review of Internet 2.0

We review the efforts to build I-2 and the progress made to date against the backdrop of the evolution of the Internet discussed above. Especially since the vision of I-2 appears to be floundering it is profitable to take stock of the ongoing efforts to build I-2 against the backdrop of a related success story. Such a comparison could help correct the course as necessary.

The first contrast between the histories of the Internet and I-2 is that nearly a dozen years after the notion of a cyber-physical infrastructure was seriously envisioned, we still do not have a *general-purpose open prototype of I-2* with a user base that is indicative of an emerging global infrastructure. As we discussed in Chapter 6, the EPCglobal Network is geared toward a narrow application—namely, supply chains—and has not emerged as a dominant general-purpose open prototype. The *ubiquitous ID* system is designed to provide a universal labeling scheme in that the *uCode* enables the labeling of not just the items that are relevant to supply chains but also any other resource at all that we choose to label. However, the *uID Architecture,* merely provides the capability to tag objects and retrieve information about them. I-2 needs not only a framework for tagging and identifying objects but more importantly a framework in which the end nodes can communicate and interact with each other. I-2 is not envisioned to be a passive infrastructure for information storage and retrieval but rather an *active infrastructure* involving dynamic interactions among the participating end nodes.

The initiatives in Europe, Japan, China, and other countries, the purpose-designed systems such as the GEOSS and the electronic toll booths, the various enabling platforms discussed in Chapter 6 are all focused on bridging the cyber and physical worlds. The ongoing efforts have given rise to several proposals for the architecture and/or roadmap for I-2. But the sobering reality is that, more than a decade later, we still *do not have a single serious general-purpose operational prototype for I-2.* The other global infrastructures mentioned above—World Wide Web, Facebook, and YouTube—started as concrete prototypes that cumulatively garnered a large user base. It is difficult to imagine that any global infrastructure—such as I-2—will emerge as a sprawling planetwide network without going through a progressive growth, starting as a prototype. Therefore, the lack of a prototype must be taken to mean that the construction of I-2 is yet to begin.

The other glaring contrast between the trajectories of the Internet and I-2, is that *commercial organizations have been involved in the development of I-2 from its inception*. If we regard the establishment of the Auto-ID Center at MIT as the commencement of the efforts to build I-2, then from the very beginning commercial organizations have participated in stewarding its growth. The Auto-ID Center, and the EPCglobal it spawned, were focused on the supply chain space [Schuster et al. 2007]. Enlisting commercial partners in the task of building a supply chain infrastructure was a natural decision. However, over the years, it appears that the mission of building an infrastructure for supply chains has gotten conflated and intertwined with the efforts to build a more general-purpose global infrastructure, namely, I-2. The unintended consequence appears to be that the growth of I-2, at least under the umbrella of EPCglobal, is not being driven exclusively in an academic environment, but is entangled with a parallel focus on supply chains that is coupled to commercial interests.

Third, while the federal agencies funded the establishment of a prototype infrastructure for the Internet, a similar federal program to build an infrastructure for I-2 has not emerged yet. With the benefit of hindsight, a *national initiative* aimed at building an infrastructure seems essential for the successful emergence of I-2. The importance of the emerging I-2 for the national strategic interests are documented in the United States [NIC 2008], Europe [Sundmaeker et al. 2010, IERC 2012], China [Inoue et al. 2011, Yan 2011], and Japan [Inoue et al. 2011, MIC1 2012; MIC2 2012], among other countries.

In addition to supporting the construction of the Internet infrastructure, federal funding also played a critical role in promoting open-architecture networking. By mandating the adoption of TCP/IP across the entire NSFNET, NSF made open-architecture networking an integral part of the nascent Internet, preemptively preventing fragmentation of the communication protocol. In contrast, I-2's landscape appears to be fragmented with multiple protocols, technologies, and architectures, with *no agency to enforce a global mandate and no initiative to promote open-architecture networking.*

The architecture of the Internet is decoupled from the applications running at the edge, which enabled the core architecture to remain simple. In contrast, *simplicity is yet to emerge in the core architectures proposed for I-2.* Implementation details of activities occurring at the edge of the infrastructure, such as whether the data from the real world is collected by devices or humans—details that are really irrelevant to the core architecture—often spill into discussions about the core architecture of I-2 [IoT-A 2011].

Considerations about *security*, which were sidelined during the incubation of the Internet, often take center stage in the discussions about the new I-2 paradigm even before a prototype has been constructed. Arguably, security is an important aspect of any global infrastructure, and currently a significant amount of attention is devoted to security on the Internet. However, as Vint Cerf remarked [Cerf 2010], a heavy focus on security in the early phases would have likely hindered the construction of the Internet. Again, the security issues could be sidelined during the incubation of the Internet since it was built within the academic research environment.

Unlike the fledgling Internet, which had *killer applications*—email and the World Wide Web—that enticed people to use the Internet, the progress on I-2 has likely suffered due to the lack of an enticing killer application. To date, there is no widely used application that makes the case for I-2. Creating such an application could be a game changer for the evolution of I-2.

Finally, from the early days of its growth the Internet exploited existing infrastructures such as the telephone lines, satellite networks, and cable networks to provide enhanced connectivity to users, while imposing minimal economic overheads. Exploiting the existing infrastructure lowered the cost barrier for connecting to the Internet and promoted an expanded user base. Although I-2 has access to similar opportunities, in particular, the vast population of smart phones, the efforts to build I-2 do not appear to have exploited the existing infrastructure as effectively.

Summary

The successful global infrastructures, such as the Internet, web, Facebook, and YouTube, started as concrete operational prototypes that were unmistakably the embryonic versions of the global infrastructures that they would later become. The importance of starting the construction of a global infrastructure with a seed prototype is self-evident and difficult to overstate. The progress on building I-2 has been hampered by the lack of a general-purpose operational prototype. The absence of a concrete prototype with a serious, accreting user base, signals that the construction of the I-2 infrastructure is yet to begin. Further, it also means that the proposed architectures for I-2 have to be regarded as *open loop designs* that have not been vetted within the framework of a large operational prototype. Hence, the critical first steps in building I-2 are to converge on a preliminary architecture and use it to build a widely deployed operational prototype. Accordingly, in the following chapters we focus on the architecture and the construction of the prototype.

8

Design Lessons from the Internet and the Web

Following the discussion of the history of the Internet in the previous chapter we review the design principles embodied in the Internet and the web. The following discussion overlaps partly with that in the previous chapter. The design principles reviewed below provide valuable guidelines for architecting global infrastructures.

The design philosophy embodied in an infrastructure is best articulated in the words of its creators. Accordingly, we have interwoven the discussion of the design principles of the Internet and the web with excerpts from the writings of the very people who formulated them, to the extent possible. The following discussion assumes an understanding of the architectures of the Internet and the web, discussed in Chapters 3 and 4.

Following a discussion of the design principles that underlie the Internet and the web, we present an overview of selected software architecture styles, culminating in a discussion of the REST (REpresentational State Transfer) style, which underlies the web. The REST style sharpens the focus on the characteristics that are desirable in an Internet-scale infrastructure. The principles outlined in this chapter provide an essential backdrop for the discussion of the architecture of I-2 in Chapter 10.

The Internet and the Web: Design Principles

The Internet and the web are distributed infrastructures that are not owned, controlled, or promoted by any single organization. Without advertising or marketing efforts to promote their use, spurred instead by demand, the infrastructures have come to pervade the whole globe, transcending national boundaries, and cultural and linguistic differences. The applications running on the infrastructures and the devices that connect to them have become increasingly diverse and demanding. The numbers of users and devices connecting to the infrastructures have increased dramatically since their inception. And yet the infrastructures have been able to accommodate the staggering growth in diversity, demand, and size without any essential change in their architectures and without any degradation in their performance. They are remarkably robust and resilient. These impressive characteristics are shared by both the Internet and the web. Their success and the similarities in their design philosophies suggest that they embody the design principles that are key to building successful global infrastructures. In the following paragraphs we attempt to identify their common design principles. The discussion is complemented with excerpts from the writings of their architects.

Universality: Both the Internet and the web are *universal platforms*. They are not *purpose-designed* for any particular type of device, network, resource, or application. Nor do they place selective barriers for devices, networks, resources, platforms, or applications that seek to access the infrastructures.

The universality of the Internet is best expressed in the words of its architects [Leiner et al. 1997]:

> *"There were other applications proposed in the early days of the Internet, including packet-based voice communication (the precursor of Internet telephony), various models of file and disk sharing, and early worm programs that showed the concept of agents (and, of course, viruses). A key concept of the Internet is that it was not designed for just one application, but as a general infrastructure on which new applications could be conceived, as illustrated later by the emergence of the World Wide Web. It is the general purpose nature of the service provided by TCP and IP that makes this possible. ...*
>
> *Beginning with the first three networks (ARPANET, Packet Radio, and Packet Satellite) and their initial research communities, the experimental environment has grown to incorporate essentially every form of network and a very broad-based research and development community."*

The paradigm of an Internet that would be interoperable with whatever networks the users chose at the edge was called the *open-architecture networking*. Its philosophy is articulated in the following excerpt [Leiner et al. 1997]:

> *"The Internet was based on the idea that there would be multiple independent networks of rather arbitrary design, beginning with the ARPANET as the pioneering packet switching network, but soon to include packet satellite networks, ground-based packet radio networks, and other networks. The Internet as we now know it embodies a key underlying technical idea, namely that of open architecture networking. In this approach, the choice of any individual network technology was not dictated by a particular network architecture but rather could be selected freely by a provider and made to interwork with the other networks through a meta-level Internetworking Architecture. Up until that time there was only one general method for federating networks. This was the traditional circuit switching method where networks would interconnect at the circuit level, passing individual bits on a synchronous basis along a portion of an end-to-end circuit between a pair of end locations. Recall that Kleinrock had shown in 1961 that packet switching was a more efficient switching method. Along with packet switching, special purpose interconnection arrangements between networks were another possibility. While there were other limited ways to interconnect*

> *different networks, they required that one be used as a component of the other, rather than acting as a peer of the other in offering end-to-end service.*
>
> *In an open-architecture network, the individual networks may be separately designed and developed and each may have its own unique interface which it may offer to users and/or other providers, including other Internet providers. Each network can be designed in accordance with the specific environment and user requirements of that network. There are generally no constraints on the types of network that can be included or on their geographic scope, although certain pragmatic considerations will dictate what makes sense to offer."*

Similarly, the World Wide Web was designed to be a universal infrastructure that was to be platform-agnostic and resource-agnostic. As Berners-Lee remarks [Berners-Lee 1996] about the web,

> *"... the real world in which the technologically rich field of High Energy Physics found itself in 1980 was one of incompatible networks, disk formats, data formats, and character encoding schemes, which made any attempt to transfer information between dislike systems a daunting and generally impractical task. ...*
>
> *Its [the web's] existence marks the end of an era of frustrating and debilitating incompatibilities between computer systems."*

Among the list of the criteria that guided his design of the architecture of the web, Berners-Lee lists the following three guiding principles that made the web's architecture *noncoercive* and widely adopted [Berners-Lee 1996]:

- *"Any attempt to constrain users as a whole to the use of particular languages or operating systems was always doomed to failure.*
- *Information must be available on all platforms, including future ones.*
- *Any attempt to constrain the mental model users have of data into a given pattern was always doomed to failure ...*

> *The author's experience had been with a number of proprietary systems, systems designed by physicists, and with his own Enquire program (1980), which allowed random links, and had been personally useful, but had not been usable across a wide area network."*

Finally, the constructs like *Uniform Resource Identifier* (URI) serve as umbrella elements that encompass a vast diversity of resources on the web. The universality of the web stems in no small measure from the success of such constructs in accommodating the heterogeneity of resources.

The design philosophy of universality is compromised if the physical objects, in I-2, are treated as a different breed of resources than cyber objects, such as digital documents. Such fragmentation of the I-2 infrastructure based on the form of the resource—that is, based on whether the resource is physical or virtual—violates universality since such an infrastructure would not be *form-agnostic*. Universality is preserved in the I-2 architecture, discussed in Chapter 10, by creating a new abstraction, called *web-enabled service* and by viewing I-2 as a network of interacting *service agents* that provision and/or consume web-enabled services.

Substitutability: The architectures of the web and the Internet are decoupled from the implementation details of their interacting components. Each of the components—such as routers, gateways, web servers, and web clients—is treated as a black-box at the architectural level. The decoupling between the architecture and the implementation details of the components—the *substitutability*—makes it possible to substitute one version/implementation of the black-box with another without affecting the overall architecture. The flexibility that allows the implementations of the components to evolve independently, without affecting the overall architecture, is key to the success of the Internet and the web as distributed and scalable infrastructures. In Berners-Lee's words [Berners-Lee 1996]:

> *"Flexibility was clearly a key point. Every specification needed to ensure interoperability placed constraints on the implementation and use of the Web. Therefore, as few things should be specified as possible (minimal constraint) and those specifications which had to be made should be made independent (modularity and information hiding). The independence of specifications would allow parts of the design to be replaced while preserving the basic architecture. A test of this ability was to replace them with older specifications, and demonstrate the ability to intermix those with the new. Thus, the old FTP protocol could be intermixed with the new HTTP protocol in the address space, and conventional text documents could be intermixed with new hypertext documents.*
>
> *It is worth pointing out that this principle of minimal constraint was a major factor in the web's adoption. At any point, people needed to make minor and incremental changes to adopt the web, first as a parallel technology to existing systems, and then as the principle one. The ability to evolve from the past to the present within the general principles of architecture gives some hope that evolution into the future will be equally smooth and incremental."*

Scalability: The number of Internet users within the United States has increased from under 1% in 1990 to over 78% of the population in 2009. Even as the usage increased dramatically, the performance of the Internet has remained largely unaffected, and quite possibly has improved over the years owing to the advances in technology. The user base on the web has seen a similar increase without, again, any degradation in the web's performance. The performance of the Internet and the web are largely independent of the size of their user base owing to the scalability that has been built into their architectures by their designers.

Consensus regarding the meaning of the term *scalability* is yet to emerge [Hill 1990]. We will adopt a working definition. We say a service system is *scalable* if the system can be expanded to meet an increased demand by investing additional resources (and efforts) that are in proportion to the incremental change in the system's size. On the other hand, if the expansion of the system requires efforts/resources proportional to the size of the system the system is not scalable. A simple example serves to clarify the definition.

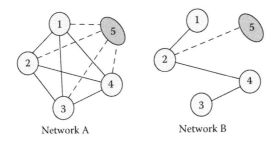

FIGURE 8.1
Illustration of a nonscalable network (left) and a scalable network (right).

Consider two local networks, say, *A* and *B*, each with four computers. Assume that in network A every computer is required to have a hard-wired connection to every other computer within the network. In network B on the other hand, assume that every computer is only required to have a hard-wired path to every other computer with the paths allowed to go through other intermediate computers in the network, as shown in Figure 8.1.

Now consider the resources/efforts needed to add a fifth computer to each of the networks. In network *A*, adding the fifth computer involves deploying four additional hard-wired links between the new computer 5, and each of the existing computers. In other words the resources/efforts needed to increase the size of the network depends on the size of the old network (that is, four links). Hence, the architecture of network *A* is not scalable, per the above definition. On the other hand adding a fifth computer to network *B* involves deploying only one hard-wired link to any of the existing computers in the network. Thus, the efforts/resources needed to grow network *B* are proportional to the incremental change in size (namely, 1), and not to the size of the old network (namely, 4). Hence, the architecture of network *B* is scalable. The growth of a scalable system is not impeded by its size. On the other hand in nonscalable systems the expansion of the system is impeded by the rising marginal cost of efforts/resources required for incremental growth. Hence, nonscalable systems present increasingly high barriers for expansion and are not suitable models for Internet-scale deployments.

Scalability of the Internet and the web are a result of the design decisions that were made during their incubation. The design principle is articulated explicitly by Berners-Lee [Berners-Lee 1996].

> *"If two sets of users started to use the system independently, to make a link from one system to another should be an incremental effort, not requiring unscalable operations such as the merging of link databases....*
> *When the web was designed, the fact that anyone could start a server, and it could run happily on the Internet without regard to registration with any central authority or with the number of other HTTP servers which others might be running was seen as a key property, which enabled it to scale."*

Per the above definition, the Internet and the web are based on scalable architectures. Creating a new resource on the web does not involve an investment proportional to the size of the web. Similarly, adding new IP-nodes to the internet involves efforts that are proportional to the size of the incremental change, and not the current size of the Internet.

In summary, an important lesson embodied in the architectures of the web and the Internet is that the barrier for expansion of an infrastructure must be kept as low as possible, if it is to support rapid user-driven growth.

Endurance: An architecture is an abstract design that is, in principle, independent of the technology used to realize it. Feasibility considerations constrain the designers to be mindful of what can be built given the limitations of the technology. However, one of the important requirements of a good architecture is that its core design must *endure* even as the technology used to realize it evolves. Both the Internet and the web have displayed such endurance. In the words of the architects of the Internet [Leiner et al. 1997]:

> "It [the Internet] was conceived in the era of time-sharing, but has survived into the era of personal computers, client-server and peer-to-peer computing, and the network computer. It was designed before LANs existed, but has accommodated that new network technology, as well as the more recent ATM and frame switched services. It was envisioned as supporting a range of functions from file sharing and remote login to resource sharing and collaboration, and has spawned electronic mail and more recently the World Wide Web."

The Internet's architecture has endured even as the technology has evolved, showing that the architecture is not coupled to any particular technology.

The technology-independence of the architecture is a particularly important criterion for I-2. Many of the objects and devices that interact with the cyber infrastructure are resource-constrained, at present. For example, a passive RFID tag harvests barely enough power from the incident radiation to run its circuitry and broadcast its serial number to an interrogating reader. Many of the smart devices are constrained by their battery life. The power requirements are playing a major role in matters such as the choice of the communication protocols and the self-organization of devices in ad hoc networks. Although the power requirements of devices is an important design consideration, if the architecture of I-2 is to endure its design should not be influenced by the shortcomings of the current technologies. The architecture itself needs to be *technology-agnostic* within the limits of feasibility.

Simplicity: All of the heterogeneity in the Internet—the different types of local networks and devices that connect to it, the communication protocols used by these devices and the local networks, their software environments— is restricted to the edge of the Internet. The core of the Internet—an infrastructure of routers and communication links that transport IP datagrams, using TCP/IP protocol—is kept minimalistic and simple. The Internet does not care about the types of networks, devices, and software environments operating at its edge. It does not care about the meaning of the data in the datagrams that flow across it. The success of the Internet's architecture is tied, in no small measure, to its inherent enduring simplicity, even as the edge of the Internet evolves to support an increasingly diverse, rich set of applications, devices, and networks. The emphasis on simplicity is contained in one of the four guiding principles Kahn used in architecting the Internet [Leiner et al. 1997]:

> *"Black boxes would be used to connect the networks; these would later be called gateways and routers. There would be no information retained by the gateways about the individual flows of packets passing through them, thereby keeping them simple and avoiding complicated adaptation and recovery from various failure modes."*

Similarly, the web also embodies an intrinsic simplicity at its core, while confining the diversity to the edge. From the perspective of the core all of the different types of digital objects are viewed as different instances of a single umbrella construct called the *resource*. The core of the web then views itself as merely a simple resource transport service. It provides a universal protocol for the transportation of resources among the end nodes of the network, not caring either about the diversity of resources or their semantics. The umbrella construct does not place any constraints on what a resource can be, allowing the complexity and diversity of resource space to evolve, even as the core architecture of the web remains unchanged. In summary, the simplicity is achieved by decoupling the core architecture of the web from the evolving diversity at the edge, using the all-encompassing construct—the resource.

The lesson from the Internet and the web is that the I-2 must function as a simple enabling universal platform if it is to succeed. The core of I-2 must be kept simple, pushing the complexity and diversity, to the extent possible, to the edge of the infrastructure.

Statelessness: The service providers on the Internet and the web, such as the routers and web servers, embody the design principle of stateless interactions. A router does not store any state information about the datagrams it routes. A web server expects the client to provide all of the required information in each interaction. The memory of previous client–server interactions is not retained by a server.

Apart from simplifying the core architecture, statelessness also promotes easy recovery from crashes, making the infrastructure more robust. The resulting robustness of the Internet and the web contain a lesson about the importance of ensuring that the transactions in I-2 embody statelessness.

Open Protocols and Requests for Comments: From the beginning the Internet and the web were based on open protocols. That is, the protocols such as TCP/IP and HTTP were distributed freely and were not owned by any commercial organization. Second, the evolution of the infrastructures was driven by a culture of open collaboration, which involved interested people working in groups and publishing their recommendations as Requests for Comments (RFCs) [Leiner et al. 1997].

The open protocols lower the economic barrier and facilitate a more rapid adoption. The inclusive and transparent process for driving the evolution of the infrastructures ensures that the refinements are driven by the long-term interests of the infrastructures, and not by the immediate commercial interests of any organization.

The next two features were discussed by Fielding [Fielding 2000] in the context of the web, but are embodied in the Internet infrastructure as well. They are desirable features for any infrastructure that seeks to enjoy global presence and sustained growth.

Low Barrier for Entry: The web grew by voluntary participation. The early adopters of the web populated it with content, converting the existing documents into hypermedia documents and posting them on the web. The barrier for conversion to hypermedia was lowered by the development of HTML, while the barrier for hosting the documents on servers was lowered by the standard HTTP that early adopters could use. At the client side the development of user-friendly browsers, and subsequently search engines were key developments that lowered the barrier for accessing content and searching for content, respectively. Similarly, the development and deployment of the TCP/IP stack, and the use of existing telephone networks initially lowered the barrier for independent networks to connect to the fledgling Internet.

Voluntary participation is also the most viable growth model for I-2. Therefore, the prototype infrastructure for I-2 must embody a low entry barrier if the infrastructure is to gain foothold worldwide.

Low Barrier for Expansion: The expansion of the Internet can be separated into the expansion of the core and that of the edge. For example, the expansion at the very edge could involve adding nodes to a local network behind a gateway. The Internet architecture allows for such expansion of the local network without necessitating an Internet-wide propagation of the change. The core of the Internet also permits expansion of both the bandwidth and the router network without requiring an Internet-wide broadcast of the changes. And yet, the newly added nodes are seamlessly integrated into the Internet ecosystem.

The web also is structured to permit independent expansion at different regions of the infrastructure. For example, to add new contents (e.g., web pages) to the web server it is sufficient to make changes within the web server. The search engine service shoulders the task of discovering the new contents and making them visible across the web.

The features discussed above embody the key design principles of the Internet and the web. Having contributed to the successful worldwide diffusion of the Internet and the web technologies they serve as useful guidelines for architecting the fledgling I-2 infrastructure. Next, we turn to an overview of a few software architecture styles. These styles provide a backdrop for the design of I-2.

Software Architecture Styles

A *software architecture style* is an abstract template for software infrastructure. Over the years several styles have been identified based on recurring patterns in software infrastructures. For an exhaustive discussion of the styles the reader is referred to Meier et al. [2009], Garlan and Shaw [1994]. The architecture styles that are germane to our discussion are outlined below.

Pipe and Filter Style: The components of the pipe and filter architectural style are software agents—called *filters*—that transform a set of inputs to a set of outputs, and connectors—called *pipes*—that route data between filters. Filters with only output ports are called *pumps*. Pumps generate data. And the filters with only input ports are called *sinks.* The implementation of the filters and pipes vary by application and involve details such as data buffering by pipes, and whether the filters operate in data-push mode at their output ports, and/or data-pull mode at their input ports [Buschmann et al. 1996, Fielding 2000]. Popularized by the UNIX operating system, the style is also illustrated in the recent Yahoo Pipes application [Yahoo 2012]. The style is particularly apt for software systems, such as compilers, in which data (source code) flows sequentially through a set of software agents that process the data [Garlan and Shaw 1994].

Client/Server Style: The client/server style partitions the interacting agents in the architecture into two classes—*clients* and *servers.* Clients are applications that generate requests, and servers the applications that process requests. For example, in the World Wide Web, which is based on the client/server style, a client would request a web page from a web server, which processes the client's request.

The clients and servers in a client/server architecture style are complemented with two types of intermediaries—the *proxy* and the *reverse proxy*—as shown in Figure 8.2. The proxy serves several clients, processing their requests and routing them to the destination

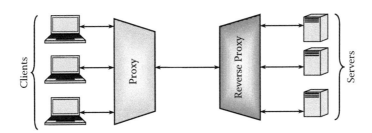

FIGURE 8.2
Illustration of proxy and reverse proxy intermediaries.

servers. The proxy could perform such services as user authentication or language translation to lighten the load on the destination servers.

While a proxy serves a group of clients, a reverse proxy serves a group of servers. Requests from either a client or a proxy could be routed either directly to a server or to a reverse proxy, which in turn routes the incoming requests to the servers it serves. A reverse proxy performs such tasks as load balancing among its servers. The responses from the servers are routed either through a reverse proxy or directly to either a proxy or a client [Barish and Obraczke 2000, Kopparapu 2012].

Variants of the client/server style include the Client-Queue-Client Style and Peer-to-Peer Style. In the Client-Queue-Client style, the server acts as a passive queue that stores the data submitted by clients. Clients use the queue to distribute data to fellow clients and synchronize data across clients. In the Peer-to-Peer (P2P) style, the agents function both as clients and servers. An example of widely used P2P architecture style was the Napster infrastructure for sharing music [Steinmetz and Wehrle 2005].

Component Style: The component style is based on the view that a software architecture is an assembly of interacting *component modules* [Niekamp 2012]. Just as a hardware system is built using components, such as resistors, capacitors, power sources, and integrated circuits, a software system with a component-based architecture style is built using reusable software components. The components are black boxes that present two types of interfaces—the *provided interface,* which exposes the services offered by the component, and the *required interface* through which the component receives the services it needs from other components, as shown in Figure 8.3. A provided interface may pertain to an offered service or could be a channel for event notification. Similarly, a required interface might

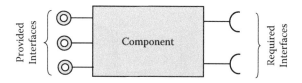

FIGURE 8.3
The structure of a component. The services provided by the component are exposed through the provided interface, while the services that the component requires from the other components are received through the required interfaces.

correspond to either a sought service or a channel to receive an event notification.

A key feature of a component is *encapsulation*. That is, the internal details of a component, including the implementation details of the services it offers are hidden inside the component and cannot be accessed by other components.

The software inside two different components could be written in different computer languages. They might even run on different machines and operating systems. In order to cobble heterogeneous components together one needs a software development environment that supports interoperability of components. Several such software development environments have been developed. An example of such an environment is CORBA (Common Object Request Broker Architecture) Component Model. Figure 8.4 illustrates the architecture of a generic software development environment. The architecture is similar to that underlying CORBA, although generic terms are used to describe the architecture.

The *software development environment* (SDE) shown in Figure 8.4 could span multiple networks over the Internet. The SDE provides a platform over which different components can interact with each other. Components A and B, shown in Figure 8.4, could involve software written in two different languages. The interfaces offered by the two components could be mutually incompatible. The stubs provided by the SDE, called *adapters* in Figure 8.4, help components communicate with each other, even if they have incompatible interfaces. For example, consider a scenario in which Component B invokes a service offered by Component A. The invocation requires the SDE to transmit a message, involving the details of the invocation, from Component B to Component A. First, the SDE uses the adapter at B's end to translate B's message into a standard format. The message is then transmitted across the network to Component A. Before

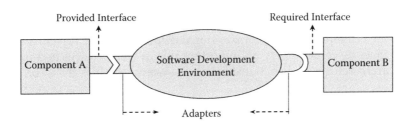

FIGURE 8.4
The Software Development Environment that facilitates interoperability of components.

delivering the message to Component A, the adapter at A's end translates the message from the standard format into the format that conforms to A's interface. By using a common description language and adapters, the SDE enables communication among heterogeneous components.[*]

Object-Oriented Style: The interacting units in this style are *objects*, which can be thought of as independent *modules* that carry their own data wrapped with software modules, called *methods*, that facilitate the access to and transformation of the data housed in the object. The objects interact with each other by passing messages. For example, the different accounts in a bank can be thought of as different objects within the bank's software infrastructure. Each account object houses data that is specific to the account, and provides a set of methods that enable another object to access the data (e.g., perform a balance inquiry) and/or manipulate it (e.g., perform a transfer into the account). In object-oriented style, primacy is given to the data housed in the objects with the methods tied to the data viewed as services to facilitate the interaction with the data. An object offers interfaces, which enable other objects to interact with it, while hiding the internal details about the data structure and method implementations from the external world.

Service-Oriented Style: In contrast to object-oriented style, in which primacy is given to the data housed in an object, in service-oriented style primacy is given to units of functionality called *services*. In a service-oriented architecture, which we will discuss in greater length in the next chapter, the interacting units are service agents that provision and consume services. A service agent also encapsulates its internal implementation details, providing only interfaces through which other service agents can interact with it. Service agents represent a higher level of abstraction in the sense that subsume the notion of objects in the object oriented style.

Other architecture styles, besides those described above, have also been identified. They include the *layered style, message bus style* and the *n-tier style*. The reader is referred to Microsoft [2009] and the references therein for additional details. Next, we turn to a style that has come to play an important role in web services and in the architectures of distributed software systems, such as the web.

---

[*] CORBA has separate adapters for the major programming languages such as Java, C, C++, Python. The standard message format used in CORBA is specified by CORBA's *Interface Definition Language* (IDL). The adapter maps the constructs in IDL to the corresponding constructs in the particular programming language used in a component [Siegel 2000].

REpresentational State Transfer (REST)

The REST style was proposed by Roy Fielding, one of the authors of HTTP 1.0 and 1.1, as a candidate architecture style for Internet-scale distributed software systems [Fielding 2000, Fielding et al. 1999, Berners-Lee et al. 1996].

Before proceeding to a discussion of the REST style, it is helpful to clarify the semantics of the terms that will appear in the discussion. The REST architecture style has abstract constructs called *components* that are interconnected using abstract constructs called *connectors*. The network constructed using the components and connectors support the flow of data that is described using abstract constructs called *data elements.*

Components are the nodes of the architectural network. The components store, provision, request, consume, and/or process information. The connectors are interfaces that help connect the components and transmit data between components. For example, a request for information is initiated by a component that is generically called a *user agent.* A web browser is an example of a user agent. A user agent seeking to send a request to a server invokes a *client* connector to transmit the request. The client connector implements the communication protocols (such as HTTP) used in the architecture and helps package and transmit the user agent's request to the server. Further, when the server sends a response to the user agent, the client connector also helps receive the response from the server on behalf of the user agent. Like a client connector, a *server* connector also receives and transmits messages. The difference between client and server connectors is that the client connector initiates requests, while the server connector operates in listen mode waiting for a request. Upon receiving a request the server connector processes the request, forwarding it to its component for processing. An *origin server* is the repository of information and is usually the endpoint of requests from user agents. The web servers that host web pages and documents are examples of origin server. An origin server responds to requests by sending the requested digital object to the user agent that initiated the request.

Intermediary components such as a proxy server could use a server connector to interface with the user agent's client connector at one end, and use a client connector to interface with a server connector at the other end, as shown in Figure 8.5. Fielding [Fielding 2000] also includes, other connectors such as *cache, resolver,* and *tunnel* connectors in the REST framework.

While the components and connectors provide abstract building blocks of the architectural network, the *data elements* provide abstractions pertaining to the information residing on and flowing across the network. We discuss three types of data elements—*resource, identifier,* and *representation.* The reader is referred to [Fielding 2000] for a discussion of the other types of data elements, namely, *resource metadata, representation metadata,* and *control data.*

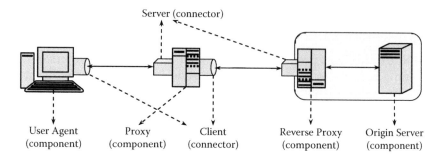

FIGURE 8.5
Components and connectors.

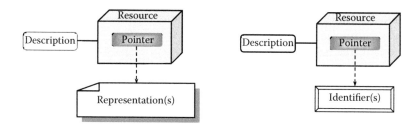

FIGURE 8.6
A resource, its description, representation, and identifier.

The critical abstraction in the REST style is the construct called the *resource*. The notion is best explained with an example. Consider the entity called "the latest published book pertaining to the Internet of Things." The quoted text is not a resource, but a *description** of a resource. A *representation* of the resource would be, for example, an electronic version of the book stored on Purdue University's library server. A representation is a concrete entity such as an electronic document. A *resource* can be thought of as a bin that contains a pointer to either a representation (or a set of representations) of a resource or to identifier(s) of the resource as shown in Figure 8.6.

The separation between a resource and its representation enables the resource to remain unchanged even as its representation changes. Thus, when a new book on the Internet of Things is published, the pointer can be redirected to point to the electronic version of the new book in Purdue's library. A client seeking the latest book on the Internet of Things will continue to request the same resource. However, what is returned in response to the request depends on what the pointer is mapping the resource to at the instant the client seeks the resource. Although the resource is a more

---

* Resource descriptions are written using languages such as the *RDF (Resource Description Framework)* language [W3C 2004].

general notion encompassing nondigital entities as well, in the remainder of this chapter we restrict attention only to digital objects.

The representation(s) of a resource are usually stored on an *origin server*. A representation of a resource typically has data contained in the resource as well as metadata about the data. For example, a file sent by an origin server could contain metadata, indicating that the *media type* of the file is "text/html," which tells the receiving user agent that the file must be processed as an HTML document. The format in which the origin server stores the resource representation is hidden from the user agent. The details such as the format in which the resource representation is to be sent can be determined by the user agent and the origin server through *content negotiation*. For example, in its request the user agent can specify the format (e.g., mpeg, mp4, quicktime, windows media video, etc.) in which it prefers to receive a requested video resource. The origin server may store the video resource in different representations corresponding to the different file formats. Or the origin server may hold the video resource in some format that is hidden, and convert it to the format desired by the user agent prior to transmission. Such details, however, are hidden behind the server connector which provides a uniform interface to client connectors. The decoupling between a uniform interface, provided by the server connector, and the origin server implementation that it hides enables the origin server to evolve without affecting the overall infrastructure.

An *identifier* for a resource is a unique label for the resource. If the resource is a book, for example, then the International Standard Book Number (ISBN) of the book is a globally unique identifier. Alternatively, the complete bibliographic information about the book comprising information about the author(s), title, publisher, edition, and publication date also serves as a globally unique identifier. Either of the identifiers—ISBN or bibliographic information—uniquely labels the book.[*]

The drawback of identifiers like the ISBN or the bibliographic information is that they are specific to books. They would not be useful for other types of resources such as, say, a video clip on YouTube. Using different types of identifiers for different classes of resources—for example, ISBN numbers for books and web addresses for video clips on YouTube—fragments the identifier space, making the architecture both cumbersome and nonuniversal. A construct called the *Uniform Resource Identifier (URI)* has been devised to provide a universal identifier space. In the URI scheme, identifiers are assigned to resources by agents with naming authority (typically the resource owners). Although agents with naming authority do not consult one another before assigning identifiers to their resources, all the identifiers are rendered

---

[*] Fielding [Fielding 2000] defines a resource more generally as a mapping from a description to either a representation or an identifier. For simplicity, we have used a more restricted definition.

unique by adding an agent-specific prefix to the identifiers. Specifically, the URI has the following syntax.

<scheme name>:<hierarchical part> [?<query>] [#<fragment>]

The query and fragment parts are optional. The following two examples clarify the syntax. Consider the following URI corresponding to a video clip on Youtube.

http://www.youtube.com/user/purdueuniversity?feature=results_main

In the above example, the scheme name is "http." The hierarchical part includes the agent-specific prefix "//www.youtube.com/" as well as the path to the resource, namely "user/purdueuniversity." Finally the query is "feature=results_main." A query is generally a sequence of (key, value) pairs separated by either a ; or an &. For example, "feature" is the key and "results_main" its value, in the above example. As a second example, consider the following URI corresponding to Isaac Newton's book *Philosophiae Naturalis Principia Mathematica*.

http://babel.hathitrust.org/cgi/pt?id=chi.11717243;seq=5;view=1up

In this URI, part of the book's unique identifier within the naming authority is passed within the query in the key-value pair *"id=chi.11717243"*. Again, *"/cgi/pt"* is the path to the resource. The other key-value pairs in the query provide metadata to be used in processing the request.

The two examples show that the URI syntax is able to reference different types of resources, such as books and videos, using a common format. However, both of these URIs are actually *Uniform Resource Locators (URLs)*. That is, each of the above examples of URI is an address to the location of the resource on the web. The URL works well if the resource one is trying to identify is on the web. However, the URL construct was inadequate to reference objects that are not on the web, or, going further, abstract resources such as, say, an idea. After much debate the consensus has converged to the current practice, which is to continue to use the above format for URI even for referencing a nondigital resource. The URI scheme does not distinguish or care about the distinction between digital and nondigital resources. The client initiating a request using a URI also does not care about the distinction. Currently, the distinction is implemented at the origin server that receives the request. If the requested resource is a digital resource available on the origin server, then it returns a representation of the resource with a status code 200, signaling successful processing of the request. The nonavailability of the requested resource at the origin server is communicated through the returned status code. For example, in the absence of the requested resource, if the server wants to return nonauthoritative information obtained from other sources, then it could use return status code 203. Or the server could choose to redirect the request using status code 303. In any case, the URI provides a scheme that enables a client to address all types of resources—digital, physical, and abstract resources—using a uniform format [Berners-Lee et al. 1998].

The current practice of restricting the distinction between digital and nondigital resources only to the purview of the origin server provides a good segue into the Internet of Things, which deals with both digital and nondigital resources. If an origin server is to provision not only the digital resources residing in its directories but also, say, data from some physical objects connected to the server, then the enhancement requires no change in the URI scheme. The client can continue to use the same URI format for requesting a nondigital resource as well. Only the implementation at the server side needs to be changed. Instead of returning a message signaling nonsuccess and nonauthoritative/redirection information, the server must be reconfigured to return the data from the requested nondigital resource. Thus, the URI scheme has greater generality built into it than is being exploited currently.

The preceding paragraphs provide a brief overview of the components, connectors, and data elements, which appear in a discussion of the REST architecture style. The following paragraphs describe the REST style.

The REST architectural style can be derived starting with several other architecture styles and progressively imposing additional constraints. In the following discussion we have reproduced one such derivation, described by Fielding, that starts with the client/server architecture style [Fielding 2000]. The REST architecture style is derived starting with the client/server architecture style and imposing the four additional mandatory constraints and one optional constraint on the client/server style. The additional constraints are described below.

1. **Stateless Client/Server Communication:** This constraint specifies that every request sent by a client should include all the information that is necessary for the server to process the request. That is, the server is constrained not to store any information pertaining to the client between requests. While this impacts the efficiency of client-server interaction, it enhances the scalability of the architecture and its resilience.

2. **Cacheable Data at Client Side:** While the server is constrained not to store any information about the client, the client itself is allowed to cache the responses from the server that are marked cacheable. The cacheable data serves to reduce client/server communication whenever possible.

3. **Uniform Interface:** This constraint places restrictions to ensure a uniform interface between servers and clients. The objective of the uniform interface constraint is to enforce an infrastructure-wide

communication framework that is understood and used by all the components, connectors, and data elements in the architecture. Such a common framework, in a sense, provides a "common language" that enables interoperability among the different pieces of the infra-structure. The uniform interface is implemented through the following four constraints.

a. **Resource Identification:** This constraint mandates that the architecture use a uniform scheme for identifying resources. For example, the resources in the web infrastructure are identified using the URI scheme described above. Using the common URI scheme enables the web server to understand which resource the web client is requesting through a specified URI.

b. **Manipulation of Resources through Representation:** The same resource (e.g., a video clip) can exist in several allotropic forms (e.g., mpeg format or wmv format) called the "representations of the resource." Implicitly, this constraint posits a distinction between a resource and its representation. Further, it restricts the allowed actions (manipulations) on a resource to the operations on its representations allowed by the architecture.

c. **Self-Descriptive Messages:** This constraint requires the messages exchanged by the components of the architecture to be self-descriptive. All the information that is necessary to process the message and perform the actions indicated in the message is to be included in the message itself. In order to process a message, the recipient should not be expected to have or need any contextual knowledge beyond what is specified in the message. Although this constraint increases the lengths of messages and slows down the interactions among components the loss in speed is offset by the gain in robustness. This constraint also promotes scalability. Adding new components is made easier by this constraint since the added components do not need any contextual knowledge in order to function.

d. **Hypermedia as the Engine of Application State:** This constraint is best explained by taking a closer look at the interaction between a user agent and an origin server. In general, an origin server provisions elaborate services, which could involve several rounds of communication between the user agent and the origin server. For example, consider the example of a user agent inter-acting with a search engine.

The user agent initiates the session by sending a search engine's server the URI of the search engine service, such as *Google*. In response, the server returns a web page to the user—that is, a *hypermedia* document—that contains a field for entering the keywords for search, and possibly some hyperlinks—e.g., hyperlinks labeled **Images, Maps, Documents,** etc.—embedded in the web page. The user agent's interaction with the search service involves either entering keywords for search, in the provided field, or following one of the hyperlinks in the web page. For example, clicking the **Images** hyperlink changes the state of the search service application from a general search service to a search service that returns only images. Such a change in the state of the application (search service) is driven by the user agent's interaction with the hypermedia document—that is, by the user agent clicking on the **Images** hyperlink in the hypermedia document. This constraint states that the only changes in the application's state that an origin server should allow should be those that can occur as a result of the user agent's interaction with the hypermedia document sent to the user agent from the origin server. For example, assume that the hypermedia document sent by a search service does not provide an option of restricting search to audio files. Then this constraint states that it should not be possible for a user agent to send an *out-of-band* message— that is, a message outside of what is allowed by the hypermedia document sent by the server—and change the application into a state wherein it functions as a search service that returns only audio files. The user should not be allowed to make the application do what it is not intended to do. Or the hypermedia must be the only engine that the user can use to change the application's state.

In addition to the four mandatory constraints described above, the REST embodies the following optional constraint.

4. **Code-on-Demand:** This constraint states that the functionalities of a client should be restricted to the set of static functionalities built into a client, optionally extended by some dynamic functionalities provided by software modules (code-on-demand) that the client can obtain from the server. While using the code-on-demand could improve the efficiency of client server interactions, and even reduce the load on the server, the downside is that it exposes the infrastructure to nonstandard and possibly malicious code.

Summary

The architectures of the Internet and the web embody many lessons for designing global Internet-scale infrastructures. In this chapter, we have attempted to glean a set of design guidelines from the Internet and the web that could be useful in architecting I-2. Following the birth of the web, the so-called REST style has emerged as an influential architectural style for large infrastructures. The chapter contains an overview of the REST style as well. The review of the design philosophies that have been baked into the Internet and the web, and the outline of the REST style, provides the backdrop for the discussion of I-2's architecture in Chapter 10.

9

Services and Web Services

The industrial landscape can be broadly divided into three sectors. The *primary sector* comprises industries that procure raw materials (e.g., mining industry, oil, and gas extraction industry). The industries in the *secondary sector* are concerned with the manufacturing of finished products using raw materials (e.g., automobile industry, computer industry). Unlike the primary and secondary industries, which deliver tangible products, the defining characteristic of the *tertiary sector* is that industries in this sector produce intangible entities called *services*. The tertiary sector is also called the *service sector* [Weber and Burri 2013].

Over the years the service sector has come to play a dominant role in the modern economy. The service industries account for about 75% of the employment in the private sector in the United States [National Academy 2007], and about 78% of the national GDP [SIFMA 2010]. Despite being a dominant component of both the national GDP and the private sector employment, the service sector is studied less than the other parts of the economy [Spohrer et al. 2007].

The word *service* refers to a broad spectrum of activities ranging from human-centric services, such as healthcare, to non-human-centric services, such as maintenance of aircraft engines. The diversity and heterogeneity of activities that are called services make it difficult to construct an all-encompassing definition of service. In fact, *General Agreement on Trade in Services (GATS)*, an international treaty forged under the umbrella of the World Trade Organization, shied away from formulating a definition of the term "service" due to lack of consensus [Weber and Burri 2013]. Without attempting to construct an all-encompassing definition, our first task will be to identify the features shared by most services to construct a working definition of the notion of service.

The heterogeneity of services also makes the classification of services—that is, grouping of services into functionally similar categories that are *mutually exclusive and collectively exhaustive (MECE)*—a challenging task [Panik 2005]. A number of different schemes for classifying services have been designed. For example, members of GATS use the Services Sectoral Classification List, called the *W/120 list*, which groups services into categories like *business services, communication services, financial services, health and social services* and so on. For details on W/120 as well as other classification schemes the reader is referred to [Weber and Burri 2013, Dumas et al. 2003, Scheithauer and Winkler 2008]. Services can also be grouped by the platform over which they are delivered and consumed. A particularly important family of services is the category of *web services* that are delivered over the World Wide Web. The paradigm of web service is of interest since a variant of it—namely, web-enabled service—serves as the unit of transaction in the I-2 infrastructure.

Several architectures have been proposed to support the provisioning and consumption of web services. Among them two styles—the WS-* and the SOAP—have garnered considerable support. Following an overview of web services, we present an overview of these two prominent architectures. This chapter provides the background for the discussion of the architecture for I-2 in Chapter 10.

Services

As mentioned above, our first task is to construct a working definition of the term "service." We will start by considering several diverse examples of services in an attempt to identify the common denominator that underlies most services. The first example is the pervasive service infrastructure—the postal service. The postal service transports packages that conform to its guidelines. The end users—sender and recipient of a package—are the consumers of the service, and the postal infrastructure is the provider of the service. The precise implementation of the service—the details of how the package gets transported—are hidden from the end users.

Next, consider a television set. It provides the service of converting the incoming signal into an audiovisual output. Thus, a television set is also a service provider. The service it provides benefits the cable company as well as the residential customer, both of whom can be regarded as the consumers of the service. Unlike the postal service infrastructure, which relies on human agents to implement its service, the television relies entirely on electronic circuitry and has no human agents inside it.

A web browser is also a service provider. Given the address of a resource on the web a browser provides the service of retrieving the resource and displaying it in a user-friendly format. The human user interacting with the browser is the consumer of the service and benefits from the service provided. Unlike the preceding examples of service providers, which involved human agents and hardware, a browser is a software module.

Shifting from the macroscopic world to the microscopic regime, consider the reaction

carbonic acid (HCO_3^-) + proton (H^+) → carbon dioxide (CO_2) + water (H_2O)

that is catalyzed by the enzyme *carbonic anhydrase.* In the process the enzyme itself is unaltered. The enzyme provides the service of accelerating the reaction breaking down about a million carbonic acid molecules per second. The enzyme's service is critical to several marine algae. The marine algae that need carbon dioxide for photosynthesis take in carbonic acid, which is about 200 times more abundant than carbon dioxide in seawater. The enzyme is then used to rapidly convert carbonic acid into carbon dioxide [Kelly and Latzko 2006]. The enzyme ensures that the supply of carbon dioxide does not become rate-limiting for photosynthesis in marine algae. In this example, the enzyme is the service provider. The marine algae are the service consumers. The catalysis of the reaction is the service provided. And the benefit to the algae is that the enzyme helps the marine algae harvest the carbon dioxide they need from the seawater. In contrast to the preceding examples, the service provider in this example is a macromolecule that operates within the service consumer, the marine algae.

"If the bees disappear from the surface of the earth, then man would have no more than four years left to live." It is a quote that is often attributed to Albert Einstein, although the author could not find the evidence to link the quote to Einstein. Its authenticity notwithstanding, the quote highlights the service bees provide by cross-pollinating flowers. The cross-pollination is a critical service that not only helps plants reproduce but more importantly it helps generate genetic diversity. The bees are Nature's service providers. The services they provide are consumed by the plants.

A sensor device such as a thermostat is also a service provider. It senses the ambient temperature and uses the sensed temperature data to regulate the operation of the furnace and air conditioner. The furnace and the air conditioner are the immediate consumers of the thermostat's service, which helps minimize their duty cycles. The communication between the service provider and consumer in this case is through electrical or wireless signals.

The furnace and the air conditioner are not only the consumers of the service provided by the thermostat, but are themselves service providers. The service they provide is to heat/cool the air in the area they serve. The consumers of their services are the humans inhabiting the space.

Another example of a sensor device that provides a service is the cruise control system in automobiles. The service it provides is to sense the automobile's speed and use that information to maintain a constant programmed speed. The consumer of its service is the human driver, whose physical effort is lightened by the control system.

Finally, an RFID tag is a service provider. The service it provides is to report the serial number stored on it when stimulated with a radio signal. The consumer of its service is the application program that receives the serial number data from the reader. The serial number helps the application program obtain information about the tagged object. The communication between the service provider and consumer in this example is through a wireless channel.

The above examples span diverse range of agents—humans, electronic gadgets, software, molecules, sensors, and machinery. In some services, such as the postal service, a service consumer is charged a fee by the provider. In others, such as the enzyme catalysis the service provider—the macromolecule—derives no apparent benefit or receives no compensation for the service it performs. Some services, such as that performed by a thermostat, are consumed by other service providers such as the air conditioner that operate downstream in the service chain. Yet others, such as the air conditioner itself, directly benefit the end customer. Some service providers, such as a thermostat, sense the environment while others, such as the air conditioner, modulate the environment. The heterogeneous activities, though seemingly disparate, have much in common. Each of the activities has a service provider and a service consumer. The service involves a task that the provider performs. And the service activity performed by the provider benefits the consumer in some way.

The preceding examples, therefore, suggest the following working definition of a service that we will use in our discussion:

> A service is a task performed by a service provider to benefit a service consumer.

The definition has four salient components. It involves two interacting agents—a *service provider* and a *service consumer*—who could be distinct. The service could involve multiple providers and/or consumers, a possibility that has been disregarded in the above definition to keep it simple. A service is an activity, namely, the *task* performed by the provider. An activity connotes a start point and an end point in time, making time an integral part of service. Finally, and importantly, the task confers a *benefit* to the consumer. If a singer sings a song in a forest and there is no one else to hear her she is not performing a service according to this definition. On the other hand, if she sings the same song in a concert to an audience she is providing an entertainment service. A consumer and benefit to the consumer are essential components of a service.

As the preceding examples show services are being provided and consumed pervasively in the world around us. The service activities are not limited to human interactions and are widespread among plants, animals, and microbes as well. While some services entail a fee to the provider the notion of service transcends the commercial realm. However, previous classifications of services, such as the W/120 list, have a rather narrow business-oriented slant to the classification. Even the previous discussions of services have largely focused on business-related aspects such as the service contracts, service fees, security, and trust.

While business interests are important, the notion of service is far broader in scope and service is far too important a construct from the perspective of I-2 to allow it to be defined or limited by commercial considerations. Consequently, in the following discussion, we will demote the business aspects of services, such as service contracts from the central role that they have come to enjoy to a more peripheral status. The business aspects of services continue to be an important facet. Of even greater importance however is the potential role that service can play as a key construct in the I-2 architecture.

Among the various types of services we focus on one special family of services—the *web services*—whose defining characteristic is that they are provided and consumed over the Internet. We discuss the web services next as a prelude to generalizing them in the next chapter.

Web Services

A web service is a software functionality that is provided and consumed over the web. The defining characteristic of a web service is that its provider and consumer are both software agents. A simple example is a web service that returns the current share price of a stock, given the ticker symbol of the company. A web service is provisioned on the web by its provider. When invoked the web service performs a service task for the benefit of the consumer. Thus, for example, if a web service for retrieving stock prices is invoked it mines the appropriate databases and returns the sought stock price to the service consumer. The interaction between a web service and its consumer is illustrated in Figure 9.1. With a slight abuse of notation, we will not distinguish between the terms "web service" and "web service provider" in the following discussion.

The life cycle of a web service begins when its provider hosts it on a web server thereby exposing it on the web. In the past, service providers would register a newly exposed web service in a central directory (e.g., UDDI). While

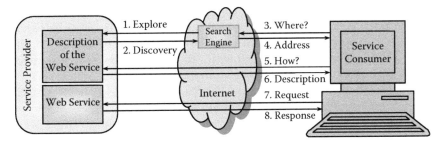

FIGURE 9.1
Schematic illustration of the consumption of a web service.

such a registration process might be viable for business services it imposes a needless barrier for expansion. Proliferation of small-scale services, such as those provided by sensors, is facilitated by transferring the burden of indexing services to an independent search-and-index utility that operates continuously. Such a utility is denoted as a search engine in Figure 9.1. The search engine periodically explores the web for newly exposed services (1), and upon discovering new services indexes them (2). A service consumer looking for a web service—say, a stock price service—queries the search engine for the desired service using appropriate key words (3). In response, the search engine returns the address (a URI) of the web service (4). Knowing the location of the service on the web does not provide a potential consumer the details about the procedure that must be followed to invoke the service. A well-designed web service, however, does (and should) come bundled with a description of itself and an associated meta-service which, when invoked, (5) provides the description of the main service to a potential consumer (6). Using the description of the service a consumer can invoke the web service (7), in response to which the web service returns the output of its activity (8).

Web services are designed to be platform-neutral and language-neutral. That is, the web services impose no restriction on the language or the operating system environment used in the service consumer's software. Such neutrality is achieved by using standard languages and protocols for communication between the service consumer and the web service, in steps (5) through (8) above. As the consumer and the web service exchange messages, each message is packaged by the sender in some standard language and protocol. Such standard languages are typically text-based markup languages. We take a closer look at each of the components involved in the interaction shown in Figure 9.1.

Anatomy of a Web Service

The anatomy of a web service is shown in Figure 9.2. A web service is typically spread over several ports in a computer offering several operations in each port. As discussed in Chapter 4, a port is addressable by external web services on the Internet. An operation is a software module that performs a well-defined task. Data transfer to and from the software module occurs through its input and output channels. Exposing a web service to the world—that is, making it available over the Internet to potential users—involves posting the service on a web service server.

An Internet-scale web service interacts with consumer software written in diverse languages and running on diverse operating systems. Such broad interoperability requires *loose coupling* between the web service and the service consumers. Loose coupling is achieved by ensuring that the implementations of the web service and the service consumer software do not interact

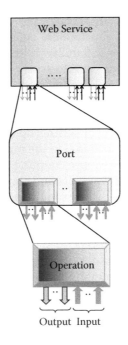

FIGURE 9.2
Anatomy of a web service. A web service spans several ports, each of which offers several individual operations. Each operation has several channels for accepting the input data and output channels through which it returns the results. A port is addressable by external users or services.

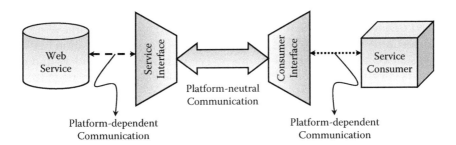

FIGURE 9.3
Loose coupling between the service provider and the consumer.

directly through procedure calls. Rather, the implementations are hidden behind interfaces. The communication between a consumer (client) interface and a web service (server) interface relies on standard platform-neutral language and protocol. Figure 9.3 is a simplified schematic of an interaction between a web service and its consumer.

The intervening interfaces between a service provider and consumer makes the interaction less efficient than it could be if the codes in the consumer and provider interacted directly through procedure calls. However, the loss in communication efficiency due to loose coupling is compensated by gains in a more important metric—scalability. With loose coupling a service provider (or consumer) does not have to worry about the language/platform used at the other end of the communication. Each agent focuses on merely encoding the outgoing messages in a standard language and protocol, and decoding the incoming messages from the standard language and protocol. Therefore, the efforts involved in setting up end nodes—service providers or consumers—is made proportional to the number of nodes added and not to the size of the network. The loose coupling also permits the independent evolution of the web service providers and consumers.

The key to a loose coupling is the language-neutral and platform-neutral communication between interfaces. The communication neutrality is achieved by using a universal data markup language, such as XML and a universal service invocation protocol, such as SOAP. For concreteness, we present a brief overview of the data markup language XML and an example that illustrates its role in achieving loose coupling. The SOAP protocol is discussed later along with the RESTful web services. For a more detailed discussion of web services see [Alonso et al. 2003, Cerami 2002, Oh et al. 2008].

Data Markup Language and Coupling

An example of a well-known markup language is HTML, which is used in hypertext documents. Similarly, a text-based language called the *Extensible Markup Language*, or XML for short, recommended by W3C (World Wide Web Consortium), has emerged as a popular choice for communication between web service providers and consumers. The web service providers and consumers use the framework of XML to encode the data that they need to send each other.

An XML document represents its data as a tree. Figure 9.4 shows a sample XML document and the associated tree. The names of elements, such as *message, to, from, user, pwd,* can be chosen arbitrarily. When data are to be exchanged it is usually transmitted by encapsulating it inside an XML file.

In order to appreciate the value of a universal language, such as XML, consider a scenario in which a retailer transacts with multiple manufacturers receiving RFID-tagged goods from all of them. Let us also assume in this toy scenario that as a service to the retailer each manufacturer offers an application called EXP_DATES_LOOKUP, which, given a list of the manufacturer's tag IDs as input, returns the expiration dates of the products corresponding to the tags. EXP_DATES_LOOKUP involves accessing

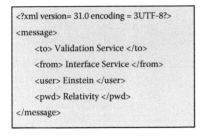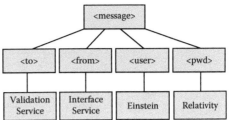

FIGURE 9.4
An XML document and its associated tree. The nodes of an XML tree are called elements. Each element (such as <user>) has some value associated with it (such as Einstein). In addition to values the elements could also have some associated attributes. Every XML document has a root node, which is labeled <message> in this example.

the manufacturer's database. The implementations of the EXP_DATES_ LOOKUP could vary among manufacturers. However, the implementation details of EXP_DATES_LOOKUP are not relevant to a retailer whose interest is confined to its input-output behavior.

Now consider an application called GET_EXP_DATES running on the retailer's data processing system. It groups the tag IDs by manufacturer and invokes the EXP_DATES_LOOKUP application for each of the manufacturers in order to compile the list of expiration dates for all the items in the retailer's inventory.

In such a scenario, consider two manufacturers who manufacture, say, bread and chips. The bread manufacturer's EXP_DATES_LOOKUP application might be written to expect the input data as a linked list, whereas the EXP_DATES_LOOKUP application of the chips manufacturer, written by a completely different developer, could be coded to expect the input data as a sorted array. Then, the GET_EXP_DATES application at the retailer's end would have to customize its communication to each manufacturer as shown in Figure 9.5.

There are two difficulties with the above messaging framework. Every time a retailer adds a new manufacturer to the supplier list, the software at the retailer end may have to be modified to write an interface application that assembles the input data to the new manufacturer's EXP_DATES_LOOKUP, in a format that is assumed in the lookup application of the new manufacturer. A more serious problem arises if a manufacturer changes the EXP_ DATES_LOOKUP application, perhaps during an upgrade of the database. For example, if the chips manufacturer wants to change the expected format of its input to a linked list, then the change has to be propagated to all the retailers that are currently interacting with the application. Each one of the retailers will then have to change their interface code or risk a broken connection to the lookup utility. Such a *tight coupling* between the applications

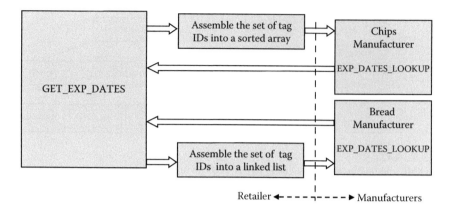

FIGURE 9.5
A tightly coupled messaging framework.

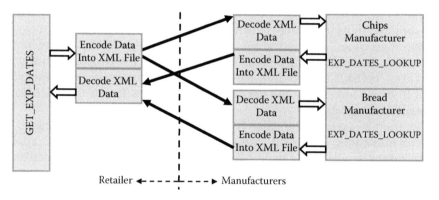

FIGURE 9.6
A loosely coupled messaging framework.

at the retailers' side and the manufacturers' side presents a high barrier for making changes at either end.

An attractive alternative would be for both the retailer and manufacturer to exchange messages using a common data markup language such as XML. The alternative messaging framework is shown in Figure 9.6.

In the messaging framework shown in Figure 9.6, the retailer converts the list of tag IDs to an XML file, without regard for the data structure (linked list or sorted array) expected by the target application. The target application, upon receiving the XML file, decodes the data contained therein and assembles the data into the data structure expected at the input channel of its EXP_DATES_LOOKUP. Thus, both the manufacturers above receive XML files. The chips manufacturer's decoding interface converts the XML data to a sorted array, while the bread manufacturer's decoding interface converts the XML data it receives to a linked list.

It is easy to see that the messaging framework shown in Figure 9.6 solves both of the problems mentioned above. Adding a new manufacturer to the list of suppliers at the retailer's end does not entail changing the code in the retailer's software. Similarly, if the chips manufacturer decides to change the data structure of the input to EXP_DATES_LOOKUP to a linked list, then all that needs to be done is to change the code in the decoding interface at the chips manufacturer. None of the retailers transacting with the chips manufacturer will be affected by the change.

Description of Web Services

Description of a service has two main purposes as shown in Figure 9.1. It facilitates discovery of the service. Second, it provides a consumer the guidance needed to invoke the service. Accordingly, the description of a service can be partitioned into two nonoverlapping components—the *discovery-oriented description* and *invocation-oriented description*. The distinction is easily understood in the context of web pages.

Description of a Web Page: The life cycle of a web page starts with the design and construction of the page as a HTML file. The HTML file contains not only the data that the page seeks to display but also metadata that provides information about the data in the file. Once an HTML file is made web-visible by hosting it on a web server, a search engine crawler discovers the existence of the page and indexes it. When a remote user interrogates the search engine, prompting it with a keyword contained in the web page, the search engine alerts the user about the existence of the web page, providing the user with the link to the page.

For example, the HTML file corresponding to the web page at www.purdue.edu contains the following metadata.

```
<meta name = keywords content = Purdue University,
Boilermakers, Boilers, West Lafayette, Indiana, United States,
college, university, higher education, academics, technology,
engineering, agriculture, health sciences, liberal arts,
libraries, research, athletics, Ross-Ade Stadium, Mackey
Arena, employment, professors, astronauts, world food prize,
nobel prize, real-world, innovation>
```

The metadata shown is not displayed on the screen by the browser. Rather, the metadata is used by search engines. When a user searches for one of the listed keywords the search engine uses the information in the above metadata to direct the user to this web page. The metadata, such as the keywords, which help a search engine facilitate discovery of this web page constitutes the *discovery-oriented description* of the web page.

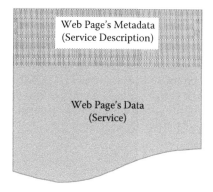

FIGURE 9.7
Structure of a web page. The web page contains the data, which comprises the information to be displayed as well as metadata such as the markup used for formatting the display and keywords that help a search engine. The metadata, that is, data about the data, describes the web page itself and is not displayed.

In addition to metadata meant for a search engine's consumption, the web page also contains metadata meant for a browser's consumption. For example, the tag

<meta http-equiv=content-type content=text/html; charset=UTF-8>

indicates that the contents found in the file are of type text and of subtype html. A browser retrieving this webpage is alerted that the material in the web page is to be interpreted as a HTML file. Further, the browser is advised to use the UTF-8 character set. Metadata such as the content-type and charset that are intended to help a browser interpret the contents properly constitutes the invocation-oriented description of the web page.

The contents of a web page may thus be logically partitioned into two parts as shown in Figure 9.7. The metadata, which is not displayed by the browser, contains the *discovery-oriented* and *invocation-oriented description* of the web page. The second component of the web page is the actual data itself, which is displayed.

The description of a web service bears considerable similarity to the description of a web page. As mentioned previously, the description of a web service contains two logical fragments that are discussed below.

Discovery-Oriented Description: This description pertains to information that facilitates discovery of services. Although keywords and verbal description, such as those included as metadata in web pages, facilitate the discovery it is possible to do better in the case of web services. Web services, unlike web pages, are relatively more homogeneous. Whereas a semantic taxonomy of the web pages is intractable a semantic taxonomy of web services seems feasible.

A *fine-grained semantic taxonomy for web services* would give a web service provider a disciplined framework for describing the category of the service being provisioned. Enabling a web service provider to annotate the provided service with a standard category to which it belongs would make the discovery process more exact. Having a web service provider annotate the service offered is also a scalable paradigm since the required efforts are proportional to the number of web services being offered.

Several classification schemes for general services are available. Examples of such classification schemes include the *United Nations Standard Products and Services Code* (UNSPSC), *North American Industry Classification System* (NAICS), *International Standard Industrial Classification of all Economic Activites Rev. 4* (ISIC) [UNSPSC 2012, NAICS 2012, ISIC 2012]. The above classifications are largely geared toward commercial services. A semantic classification scheme that is tailored for web services and encompasses both commercial and noncommercial web services is yet to emerge.

Although a semantic classification scheme for web services has not yet been developed, several frameworks have been proposed to enhance the current web service description languages, such as WSDL, to enable annotation of the WSDL-based description with semantic information. The WSDL-S recommendation by W3C and the OWL-S are two examples of such proposed frameworks [Akkiraju et al. 2005 and OWL-S 2008]. The usefulness of such frameworks would be significantly enhanced by the following two-step program.

1. Develop a fine-grained *hierarchical semantic taxonomy* for web services, representing the hierarchy as a tree. Traversing the tree from the root to the leaves should progressively narrow down the class of web services, with each leaf representing a fine-grained description of a class of web services. A prefix-based coding system should be used to capture the hierarchical structure, with common prefix indicating a shared ancestor in the tree.

2. At each leaf, that is for each narrowly defined class of web services, develop a class-specific semantic model that can be used by service providers and consumers to further describe a service within that class.

Partitioning the universe of web services using a hierarchical classification described above would enable the service providers and consumers to achieve a better semantic match.

Invocation-Oriented Description: This description pertains to information that a consumer needs to invoke the service. In contrast to the discovery-oriented description, which contains semantic information, this description is syntactic. It specifies the ports, the operations, and the input–output data formats for the operations. This description is to be communicated to the consumer in a common markup language.

Currently, the *Web Services Description Language* (WSDL), recommended by the World Wide Web Consortium (W3C), is widely used for the invocation-oriented description of web services [Chinnici et al. 2012]. WSDL is based on XML and is derived by placing some constraints on the structure of the XML file. In the following paragraphs we present a brief overview of the WSDL to illustrate how it is used to provide the invocation-oriented description of a service [W3Schools 2012]. The running example, discussed above, will be used as the backdrop for the overview.

A WSDL file contains the following major types of XML elements:

- **<types>**: specifies the data types used by the web service.
- **<portType>**: provides a description of the operations offered in a port
- **<message>**: provides a description of input and output data for each operation
- **<binding>**: provides a specification of the protocol and data format for each port

The <portType> and the <binding> are the key elements in the WSDL file. The following toy example [W3Schools 2012] illustrates the structure of the key elements using the example shown in Figure 9.6.

The <portType> element shown in Figure 9.8 describes a port named ExpirationDates. The port ExpirationDates provides an operation named ExpDatesLookup. The operation expects an input message, which is named InputData, and outputs a message named OutputResponse. The message InputData, described at the top, comprises a data item called TagID, which is a string of characters. The message OutputResponse similarly comprises a data item, ExpDate, which is also a string.

The prefix, xs: in xs:string indicates that it is a built-in data type in XML schema. (XML schemas are used to describe the structure of an XML file by imposing constraints on the structure and contents of an XML file. We will not discuss the schemas here. An interested reader could learn more about schemas by referring to the resources in W3C-Schema [2012].)

The <portType> element gives an abstract description of the port and the operations offered at the port. Such an abstract description does not tell a consumer what protocols the consumer must use to communicate with the port. The required information is provided in the <binding> element of the WSDL description [W3Schools 2012].

The <binding> element has two attributes. The type attribute is used to specify the port that the binding element is describing, which in our example is named ExpirationDates. The name attribute is used to give a name to the binding element. We have chosen the name BindingToSOAP. The web services are broadly divided, at present, into two different classes—SOAP-based web services and RESTful web services—depending on whether they use SOAP (Simple Object Access Protocol) or the standard HTTP protocol

```
<?xml version="1.0" encoding="utf-8"?>
<definitions
 xmlns:soap="http://schemas.xmlsoap.org/wsdl/soap/"
 xmlns:wsdl="http://schemas.xmlsoap.org/wsdl/"
 targetNamespace=http://www.example.com/ExpirationNameSpace/">

<message name="InputData">
 <part name="TagID" type="xs:string"/>
</message>

<message name="OutputResponse">
 <part name="ExpDate" type="xs:string"/>
</message>

<portType name="ExpirationDates">
 <operation name="ExpDatesLookup">
 <input message="InputData"/>
 <output message="OutputResponse"/>
 </operation>
</portType>

<binding type="ExpirationDates" name="BindingToSOAP">
 <soap:binding style="document"
 transport="http://schemas.xmlsoap.org/soap/http"/>
 <operation>
 <soap:operation
 soapAction="http://www.example.com/ExpDates/ExpDatesLookup/">
 <input><soap:body use="literal"/></input>
 <output><soap:body use="literal"/></output>
 </operation>
</binding>

</definitions>
```

FIGURE 9.8
WSDL description of a simple web service.

for invocation [Spies 2008]. The two invocation protocols are discussed in greater detail below. In the example shown, the WSDL file is describing a SOAP-based web service. The soap:binding element also has two attributes. The style attribute specifies whether the invocation is message-based (style = "document") or a Remote Procedure Call (style = "rpc"). The transport attribute specifies the transport protocol to use for communication, which in the above example is HTTP. The <binding> element includes an <operation> element for each of the operations exposed at the port. The soapAction attribute of the <soapOperation> element specifies the web address of the application implementing the operation; the web address includes the address of the server(www.example.com), the path to the application on the server (/ExpDates), and the name of the application (ExpDatesLookup).

The <input> and <output> elements specify the encoding of the input and output parameters, which are taken to be "literal" in the example shown.

The simple WSDL example illustrates how a web service provider furnishes the details a consumer would need to invoke the offered web service. As illustrated in Figure 9.1, after locating the web service in step 4, the consumer software requests (step 5) and obtains (step 6) a WSDL-based description of the web service. The description is then used to invoke the web service in step 7. We provide an overview of the web service invocation following a discussion of the service discovery process (steps 1 and 2 in Figure 9.1).

Discovery of Web Services

The proliferation of web pages on the World Wide Web necessitated the development of search engines that help an end user find the web pages of interest. A similar automated discovery mechanism is needed to mediate the interaction between providers and consumers of web services.

The search engines, such as *Google,* enable owners of the web pages to advertise their web pages *passively.* That is, a search engine does not expect the creators of web pages to either register or even report newly created web pages to the search engine. Instead, the search engines crawl through the web periodically discovering additions, deletions, and modifications of the contents posted on the web. The current status of the web is maintained in a search engine's database. An end user seeking to locate a web page queries a search engine's database to retrieve the links to the relevant web pages. Such a passive advertisement of the contents lowers the barrier for creating new contents. The load of discovering new contents and matching them to a consumer's requests is borne by the intermediaries—the search engines. As evidenced by the popularity of the search engines for the web, the passive advertisement is a successful, scalable model for resource discovery on the web. It is also a promising model for advertising web services.

In contrast, the *Uniform Description Discovery and Integration (UDDI)* is an initiative to build a global directory of web services [UDDI 2006; Newcomer 2002]. The UDDI infrastructure is organized as a distributed network of nodes each of which hosts three databases—called the *White Pages, Yellow Pages,* and *Green Pages.* The White Pages are designed to provide name and contact information about service providers. The Yellow Pages are designed to provide a classification of the web services, based on service taxonomies, such as UNSPSC and NAICS [UNSPSC 2012, NAICS 2012]. The Green Pages are designed to provide technical invocation details about the web services. The UDDI registries encompass a collection of UDDI nodes that host UDDI-compliant information. The UDDI framework enables service providers to register their web services in public directories, which can then be queried by service consumers.

The UDDI initiative is an example of a framework that requires *active advertisement* by the service providers. The burden of registering services that UDDI places on the providers raises the barrier for participation by web service providers. The UDDI network has suffered setbacks. IBM, Microsoft, and SAP, three of the prominent supporters of the initiative, closed their public UDDI nodes [UDDI 2006].

The search engine model of passive advertisement of web resources has been significantly more successful for resource discovery on the web than the UDDI model of proactive advertisement has been for web services. Although the resources on the web are a different class of entities than web services there is considerable overlap between the two classes. Some of the resources exposed on the web, such as the application resources that retrieve stock prices or weather information are indeed web services. A passive advertisement paradigm mediated by search engine is hence more desirable than an active advertisement paradigm for discovery of web services. Accordingly, we have chosen to illustrate a search engine and not a registry in Figure 9.1 At the present time, we do not have a widely used search engine dedicated to indexing and discovering web services.

Invocation of Web Services

Service consumption activity is illustrated schematically in Figure 9.1. The final steps in the consumption are the invocation of the web service (step 7) and reception of the output response from the service provider (step 8). Characteristic of a technology that is yet to reach maturity, there are competing protocols for service invocation. Among the invocation protocols, two have gained prominence and the web services are often divided into two classes—the SOAP-based web services and RESTful web services—based on the two invocation protocols—SOAP (Simple Object Access Protocol) and HTTP (HyperText Transfer Protocol).

The two protocols differ in the manner in which they implement the invocation of a web service. SOAP-based invocation involves sending a SOAP request, that is, a message containing a request for service, written using the SOAP conventions. A consumer prepares a SOAP request using the WSDL description of the web service retrieved from the provider. The response from the server is returned as a SOAP response, that is, as a message containing the results, also written using the SOAP conventions. In contrast, a RESTful web service is provided as a resource on the web. A RESTful web service is invoked using the same HTTP conventions that are used to access resources on the web. The results of the execution of the service are also returned as a HTTP response.

RESTful web services have the advantage of relative simplicity since they do not need an added protocol, such as, SOAP, for service invocation. The widely used HTTP serves as a basis for RESTful web service invocation as well. Although an effective language to describe RESTful web services was not available previously with the advent of WSDL 2.0, RESTful web services can be annotated with a WSDL-based description as well [Mandel 2008]. The SOAP-based web services, on the other hand, though more cumbersome offer better security features and end-to-end reliability, which are lacking in RESTful web services at present. For a detailed discussion contrasting the SOAP-based and RESTful web services, the reader is referred to [Newcomer 2002].

Summary

The web services—services provisioned and consumed by software agents over the web—provide a particularly useful template for the *web-enabled services* that serve as the key umbrella construct in I-2. In preparation for the discussion in Chapter 10, we have reviewed some aspects of services and in particular web services. The discussion in this chapter serves as a prelude to the discussion of the service oriented architecture for I-2 in Chapter 10.

10

Architectural Imperatives for I-2

The efforts to bridge the cyber and physical worlds have progressed along several directions, as described in Chapter 6. Ranging from national and international initiatives such as smart grid and IPSO Interops to the activities of hobbyists and do-it-yourselfers, from web-based control of residential appliances to monitoring individual items in supply chains, the ongoing endeavors span a broad spectrum. While the vast breadth of activities has certainly helped stimulate interest, it may have also splintered the focus of the endeavor. Despite worldwide incubation efforts for nearly a decade no single initiative has emerged as a clear candidate for, or an embryonic version of, the envisioned cyber-physical infrastructure. In fact, let alone an embryonic version of the infrastructure, at present, we do not even have consensus on a blueprint for the cyber-physical system. There is no global agreement on what the infrastructure should be, on its definition or its scope. Instead, characteristic of an endeavor that is yet to reach maturity, there is a profusion of proposals for the definition, scope, and architecture of the fledgling infrastructure.

Following the terminology used in previous chapters, we will continue to refer to the emerging cyber-physical infrastructure as Internet 2.0—or I-2—to separate it from the various models that have been labeled with the generic term "Internet of Things" in the literature [IoT-A 2011]. In this chapter we outline the architectural imperatives for I-2, by which we mean the set of features that must be baked into I-2's architecture. Our discussion of the architectural imperatives is guided by the lessons learned from the Internet and the web. The architectural imperatives are then used to provide a coarse-grained description of I-2 and its scope. The endpoint of this chapter is not a proposal for I-2's architecture. Rather, it is a set of principles that must be incorporated into I-2, if it is to emerge as a successful Internet-scale infrastructure. The discussion here will hopefully serve to facilitate convergence towards a global consensus on I-2's architecture.

The rest of the chapter is organized as follows. Viewing I-2 as a distributed complex system, we first ask and answer the question: *what is the right resolution scale for I-2?* The I-2 paradigm encompasses a vast heterogeneity of physical resources, digital resources, communication protocols, and applications. While supporting a vast heterogeneity at the edge, à la the Internet, the core architecture of I-2 must remain essentially simple and homogeneous if it is to be robust on an Internet-scale. The Internet, and the World Wide Web, have achieved such simplicity in their core architectures through a judicious choice of the resolution scale. Thus, the choice of the resolution scale is a key architectural imperative.

A choice of the resolution scale necessitates the development of a *transport protocol* to support communication at the chosen resolution scale. We outline the specifications of the transport protocol, drawing upon the discussion in Chapter 9. The transport protocol for I-2 is required to operate not only on the Internet but also in air-based machine-to-machine (M2M) communications that do not rely on the Internet.

Whereas the Internet relies on fixed routers to transport the datagrams, I-2 is expected to operate over a network comprising both stationary as well as mobile routers. While the discovery and advertisement of nodes are not essential features of the Internet, they emerge as critical new capabilities that must be incorporated into I-2's architecture. We describe the new capabilities that are needed, postponing discussion of a blueprint for their implementation to the next chapter.

Complex Systems and Resolution Scale

Complex systems typically contain interacting components that can be viewed at various levels of granularity. In architecting complex systems, the correct choice of the resolution scale—that is, the choice of the granularity at which interacting entities are to be viewed—is a critical decision. Choosing a resolution scale amounts to deciding what entities are to be viewed as the irreducible components—or *atoms*—of the system. The finer structure of the irreducible components is then ignored, and the complex system is viewed at a resolution that is no finer than that of its irreducible components. A judicious choice of the resolution scale greatly facilitates the design, analysis, and management of a complex system, even as a poor choice injects counterproductive and distracting details into the picture. The following real-world examples illustrate the importance of choosing the right resolution scale.

Consider a car mechanic engaged in repairing a car. His view of the automobile is that it is built using parts that he considers irreducible. For example, from a mechanic's perspective a nut or a bolt is an irreducible component of the automobile. A mechanic does not care about the finer structure of a nut or a bolt. From a physicist's perspective, the nut is most certainly not irreducible, since it is made of atoms, which are in turn made of electrons and nucleons. The nucleons themselves are made of quarks, held together by messenger bosons. However, to a mechanic the finer structure of the nut at atomic, nuclear and subnuclear scales is of little relevance, and in fact a counterproductive distraction. Considering the structure of a nut or a bolt at, say, atomic resolution would overwhelm his task with needless details. Instead, treating nuts and bolts as irreducible components enables him to hide the needless details, and allows him to focus only on the relevant details. All he needs to know about a nut are some gross properties such as its size, and not how the atoms are organized inside to make the nut.

To a traffic engineer, interested in streamlining the traffic, and in minimizing accidents and congestion, the most appropriate resolution scale is to consider an entire vehicle to be an irreducible unit. The internal structure of an automobile—that is, how its parts are put together—are details of little relevance to a traffic engineer. From his perspective, such details are best hidden inside what he takes to be irreducible—the vehicle. Taking a vehicle as an irreducible unit, and hiding its internal details, enables the traffic engineer to focus on the intervehicle interactions and the traffic flow patterns without having to deal with intravehicle details. Thus, a traffic engineer's resolution scale is much coarser than that of a mechanic's.

To a chemist studying chemical reactions, however, it is convenient to choose a much finer resolution scale than that of a mechanic. A chemist would take the irreducible units to be atoms. Chemical reactions, which do not involve ionization or transmutation, amount to rearranging atoms in the reactant molecules to form different product molecules. The membership of atoms in different molecules changes during a reaction, but the atoms themselves remain unaltered. Therefore, treating atoms as irreducible building blocks provides the convenient abstraction that hides the subatomic, nuclear, and subnuclear structures of atoms, while bringing into focus structures and interactions at atomic distances that are of interest to a chemist.

The above examples show the importance of looking at complex systems at the right resolution scale, and of choosing what one deems as an irreducible, indivisible unit in the system. Choosing the right granularity, the right entity as the irreducible unit and hiding the details of the structure inside that irreducible unit is a critical decision for the effective design, management, and analysis of complex systems.

As the preceding discussion suggests, an important decision in architecting I-2 is the answer to the question: *what are the atoms of I-2?* I-2 will be composed of a heterogeneous mix of entities such as physical objects, sensors, RFID tags, electronic devices, software agents, humans, computers, and routers, to mention a few. Amid this bewildering variety of interacting components, what should we choose as the irreducible units of I-2?

Web-Enabled Services

In order to answer the above question about the irreducible units of I-2, we consider constructs called *web-enabled services* and *service agents*. The constructs are described using the following hypothetical example.

Consider the task of enforcing toll payments at toll booths. If a vehicle does not pay the toll then data about the vehicle needs to be captured for follow-up action. So, the desired functionality of a surveillance mechanism that we call a *Data Acquisition Module (DAM)* is that it should transmit the license plate number of an offending vehicle to a central computing facility.

Figure 10.1 shows two different internal structures for the Data Acquisition Module. We will assume, for the sake of the following discussion, that all vehicles carry RFID transponders for electronic toll payment. In the first implementation of DAM, shown on the left in Figure 10.1, a reader in the DAM system could acquire the serial number of the transponder and interrogate the right database to retrieve the license plate number corresponding to the transponder. The input to DAM in this case would be a serial number of the transponder. The serial number would be transmitted from a transponder as radio waves, and the output would be the license plate number of the vehicle as shown on the left in Figure 10.1.

An alternative design for DAM would be to use high-speed cameras. High-speed cameras can be deployed to photograph the license plate number of a vehicle if it passes through a toll booth without paying the toll. The photograph can then be fed to character recognition software to obtain the

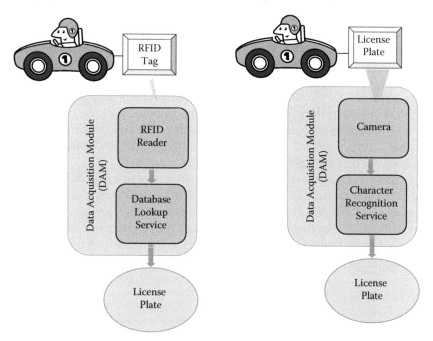

FIGURE 10.1

Two different implementations of a data acquisition system. The implementation on the left is based on RFID technology and that on the right uses an image capture mechanism and automatic character recognition. Both the implementations transmit the license plate number of the vehicle to a central computing facility.

license plate number automatically. In this case, the input to DAM is a visual image, recorded by a camera, and the output, again, is the license plate number of the offending vehicle, as shown on the right in Figure 10.1.

Figure 10.2 illustrates the architecture of a hypothetical Toll Enforcement System and the cascade of events involved. The first step in the cascade, denoted ①, is the acquisition of the data from the vehicle—using one of the two implementations of DAM shown in Figure 10.1. The license plate number is then transmitted over the Internet to a Central Computing Facility (CCF), which could be located in a different city. Using the license plate number, the computing facility determines which database it must query to obtain the details of the vehicle, such as the address of the owner and the details of the account tied to the license plate number, if any. Again, the vehicle could be registered in a different state from the one in which the infraction occurs and the database containing the information about the vehicle could be located at a different location than the central computing facility. ③ So, the computing facility transmits the request for data regarding the vehicle to the right database over the Internet. Assume that there is a software layer wrapped around the database, called Database Lookup Service (DLS), that retrieves the requested information from the database. ④ The details about the vehicle

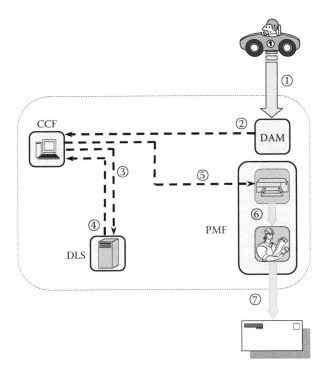

FIGURE 10.2
Architecture of a Toll Enforcement System. CCF: Central Computing Facility. DLS: Database Lookup Service. DAM: Data Acquisition Module. PMF: Print and Mail Facility.

are transmitted by the DLS to the computing facility, again over the Internet. If the owner does not have sufficient funds an electronic account for toll payment, or if the account does not have sufficient funds then the computing facility initiates a notification to the owner. ⑤ A letter containing the details of the incident is generated and sent to a Print & Mail Facility (PMF), over the Internet. ⑥ The printer generates a hard copy of the letter. ⑦ A human operator packages the letter and ships it.

The DLS and PMF units in the above architecture provide services that embody an important difference. The DLS accepts a license plate number as input and returns the associated vehicle information as output. The DLS is invoked over the Internet by a software agent (running in CCF), and the service request is fulfilled by a software agent (wrapped around the database). Therefore, DLS is a web service.

The PMF also is invoked over the Internet by a software agent running in the CCF. The PMF accepts an electronic version of a letter as input. The letter is then printed and mailed by PMF. The service provided by PMF is invoked by a software agent running in CCF. But the execution of the requested service involves, in addition to the printer's hardware and software, a human agent who does the mailing. That is, the service provided by PMF is not implemented using only software agents and is hence not a web service. We call services such as those provided by PMF, *web-enabled services*. A service is called a *web-enabled service* if it is invoked electronically, with all the resources necessary to execute the service included in the electronic invocation. The implementation of a web-enabled service is not restricted to rely only on software agents. Instead, the implementation of a web-enabled service could involve physical objects, software agents, digital resources, and human agents, as illustrated in the PMF.

Web-enabled services form a subclass of general services. For example, the service offered by the human mailing agent inside the PMF is not a web-enabled service since the printer does not invoke the agent's service by sending an electronic service request to the agent. To provide another example, the tire-rotation service offered by tire dealers is not a web-enabled service, since it cannot be invoked by sending all the resources necessary to execute the service electronically to the tire dealer. Invoking the service involves transporting a physical object—the automobile—to the dealer. On the other hand, if a tire dealer provides a price quote service for tire-rotation in response to a web-based request, then the dealer would be provisioning a web-enabled service since all the material needed to perform the service, namely, the information about the vehicle, can be transmitted electronically. The service itself—determining the price tied to the service—may be performed by a human agent at the dealer's end.

Notice that a web-enabled service is only restricted to be invoked electronically, and not necessarily over the Internet or the web. The CCF could invoke the service offered by PMF using direct wireless communication that does not rely on the Internet/web, and the service provided by the PMF would

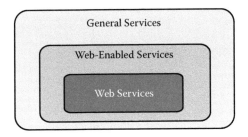

FIGURE 10.3
Containment hierarchy for services.

still be regarded as a web-enabled service. The web in the web-enabled service refers not to the World Wide Web, but to an infrastructure called the service web that is described later in this chapter. The only restriction on a web-enabled service is that its invocation must be electronic, and that all the resources needed to execute the service must be transmitted electronically. Web-enabled services place no restriction on the agents used to execute the service. Web-enabled services provide the necessary construct to integrate the cyber and physical worlds, as we show later.

Since every web service is invoked over the web, and hence electronically, every web service is also a web-enabled service. However, web-enabled services form a broader class than web services. Whereas, a web service is typically a piece of software-based functionality, the implementation of a web-enabled service could involve interaction between the cyber and physical worlds. For example, the capability to operate electrical appliances over the web is a web-enabled service since the implementation of the service, which involves embedding extra circuitry in the appliance, does not rely only on software agents. The containment relationships among the three classes of services is summarized in Figure 10.3. The remainder of the discussion in the book will pertain to web-enabled services, unless specified otherwise. With a slight abuse of notation, we will use the word service to refer to web-enabled service, hereafter.

Service Agents

The Toll Enforcement System described above has four interacting **service agents**—DAM, CCF, DLS, and PMF, as illustrated in Figure 10.2. A *service agent* is defined to be an entity that provides and/or consumes a web-enabled service. For example, the PMF provides a web-enabled service. When invoked electronically by CCF, it converts the digital file provided to it by CCF into a printed document and mails the document. Similarly, CCF provides

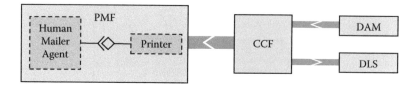

FIGURE 10.4
The Toll Enforcement System, a cyber-physical infrastructure, is illustrated as a network of interacting service agents. The service agent CCF invokes the services provisioned by PMF and DLS, and is invoked by DAM.

the web-enabled service of generating a letter using the license plate and the associated vehicle information. CCF consumes the web-enabled service provided by DAM and invokes the web-enabled services provided by DLS and PMF. See Figure 10.4.

Service agents could encapsulate a coupling to the physical world (DAM), could contain digital resources (CCF, DLS, PMF) and/or could contain human components (PMF). The definition of a service agent places no restriction on its internal implementation details.

A service agent could provide as well as consume services. For example, the CCF consumes the service offered by PMF and DLS, and in turn provides a service to the DAM as shown in Figure 10.4. The conventions used in Figure 10.4 are that the arrow points from the invoker/consumer to the provider of a web-enabled service.

A web-enabled service typically has a service provider and a service invoker/consumer. Service agents, such as DLS, that merely provide a web-enabled service and never invoke one are *pure service providers*. On the other hand, a service agent that always invokes and never provides a service is a *pure service consumer*. A typical service agent, such CCF, would be a hybrid, that provides as well as invokes web-enabled services.

Resolution Scale for I-2

The Toll Enforcement System bridges the physical world of automobiles and the cyber infrastructure of the toll booth. The cyber-physical coupling in this system is representative of the interaction between the cyber and physical worlds envisioned in I-2. Figure 10.4 shows that the architecture of the system can be described satisfactorily using the *service agents*—DAM, DLS, CCF, and PMF—as the building blocks. The question then is: should we take service agents as the irreducible building blocks of I-2, or should we use a finer resolution scale and possibly include the internal structure of the service agents, such as their implementation details, in a description of I-2's architecture?

That is, should a description of the architecture of the Toll Enforcement (cyber-physical) System include the implementation details of DAM—such as whether DAM is implemented using RFID technology or image capture technology? If DAM is initially based on, say, RFID technology and its implementation is changed to base it on image capture technology, then the change has practically no impact on the architecture shown in Figure 10.4. The image capture technology could function with the same speed and reliability as the RFID technology and offer the same functionality. Switching the DAM from RFID mode to image capture mode does not have any impact on the other three service agents (CCF, DLS, and PMF) functioning downstream either.

If changing a feature—such as the implementation of DAM—has no impact on the overall architecture, then clearly the feature in question is not a consequential detail in the architecture and should be encapsulated away. Similarly, the implementation details of the other three service agents are also irrelevant from the architectural perspective. For example, if the printer in PMF is replaced with a different comparable printer, neither the architecture nor the other service agents are impacted. Therefore, we conclude that the *implementation details of the service agents should not be included in a description of the architecture*. The resolution scale of I-2 should be chosen to be sufficiently coarse to hide the implementation details of its service agents.

In contrast to the implementation details of a service agent, the functionality of a service agent is relevant to the architecture. For example, if the functionality of DAM is changed to make it output a description of the color of the offending vehicle instead of the license plate information, then the Toll Enforcement System described previously collapses. The CCF would be unable to retrieve the necessary information about the vehicle. Since changes in the functionality of a service agent impact the viability of the whole architecture, the *functionality of a service agent is relevant to the architecture*. Therefore, the resolution scale of I-2 should be chosen to be fine enough to include the details about the functionalities of its service agents in the description of its architecture.

This discussion is consolidated into the following core design principle for I-2.

> **CORE DESIGN PRINCIPLE FOR I-2 (CDP):** The irreducible, atomic building blocks of I-2 must be *service agents* and I-2 must be viewed as a network of interacting service agents. The unit of transaction among the service agents must be a *web-enabled service*.
>
> A *service* agent's internal details are not to be included in the design or description of I-2's architecture. They must be encapsulated behind the service agent's interface and must not be visible or accessible to other service agents in the I-2 network.

To close the loop on the discussion in Chapter 8, we look at CDP against the backdrop of the lessons learned from the Internet and the web. Decoupling the architecture from the implementation details of the service agents, as suggested by the CDP, ensures that the I-2 architecture will *endure* even as the implementations of its constituent atoms *evolve independently*. For example, *substituting* a human operator with an automated packaging and mailing facility upgrades the PMF service agent without impacting the architecture. Second, the CDP imposes clarifying *simplicity* on I-2. The resolution scale chosen in the CDP ensures that the bewildering heterogeneity of entities—such as physical objects, bridge technologies, IP-enabled devices, digital resources, communication links, and human users—is encapsulated inside I-2's atoms (service agents), making I-2's architecture just a simple network of interacting service agents. Third, the CDP places no restrictions on the type of web-enabled services transacted among service agents, making the architecture *universal*. We will continue the review of CDP, against the backdrop of the discussion in Chapter 8, after presenting a description of I-2. In order to describe I-2, we need the notion of Service Transport Protocol.

Service Transport Protocol

The Core Design Principle (CDP) describes I-2 as a network of interacting service agents. For I-2 to function as an Internet-scale infrastructure, it must be based on a standard protocol that all service agents in I-2 must use to both provision and consume web-enabled services.

We recall that the World Wide Web requires all clients and servers to use a standard protocol—HyperText Transfer Protocol (HTTP)—to transport resources. All of the end nodes in the Internet are required to use the TCP/IP to transport IP datagrams. When a web-enabled service is provisioned in response to an invocation, we will suggestively say that the service is being "transported" from the service agent providing the service. Some service agents, such as DAM in Figure 10.4, may provision web-enabled services without an invocation. In such cases, we will assume that the service was provisioned in response to a *null invocation*.

As a suggestive analogy to HTTP and TCP/IP, we call the protocol for "transporting" web-enabled services in I-2, the *Service Transport Protocol (STP)*. Designing an architecture for I-2 involves designing a universal STP that governs the provisioning and consumption of web-enabled services at every service agent in the I-2 network. The STP should be designed to encompass all transactions between two service agents within the I-2 framework, including advertisement, discovery, and description of services, in addition to invocation. Further, any interaction between service agents outside the scope of STP is to be regarded as *out-of-band* and forbidden. We will discuss the specifications for STP in greater detail later in the chapter. Before returning to a discussion of STP, however, we consolidate the previous discussion to present a description of I-2.

What Is I-2?

We present a coarse-grained overview of I-2 below, deferring a more detailed discussion of its components to the sections that follow. Figure 10.5 illustrates a high-level view of I-2's architecture. As stated above, the atomic building blocks of I-2 are service agents shown as cylinders in the upper plane in the figure.

Service agents provision and consume web-enabled services. As shown in the figure, the internal components of a service agent could comprise physical and nonphysical resources as well as human components. The generality of the service agent construct provides the necessary flexibility to treat physical objects, digital resources, and human components uniformly in I-2 as components of the service agents. The flexibility enables I-2 to integrate physical and cyber worlds into a seamless infrastructure.

The interactions among the service agents involve a finite set of service-related activities such as invocation, provision, advertisement, description, and discovery, to name a few. The *Service Transfer Protocol* (STP) is to provide

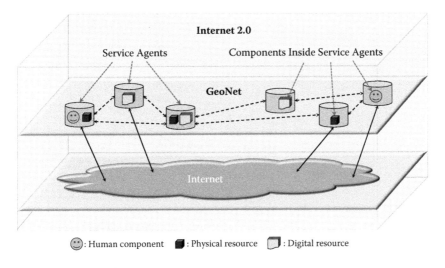

FIGURE 10.5 (See color insert.)
A coarse-grained view of I-2's architecture.

the protocols for the interactions between service agents. The STP, which is the bedrock of the I-2 architecture, is to enforce *uniform interfaces* between interacting service agents, to enable interoperability among otherwise incompatible service agents. See Figure 8.3 for an illustration of uniform interfaces.

The service agents are allowed to communicate with each other in two modes: *Internet-based communication* and *M2M communication*. The communication pathway in Internet-based communication goes through the Internet, while the pathway in an M2M communication does not. An example of M2M communication is the air-based wireless transfer of data between two Wi-Fi-enabled devices that are within each other's range. The Internet-based communication links are shown as solid lines across planes. The M2M communication links are shown as dashed lines in the top plane in Figure 10.5.

The M2M communication links are established based on geographic proximity of the service agents. Hence, the network of service agents formed by the M2M communication links will also be called a *Geographic Network* or *GeoNet* to distinguish it from the networking of the agents that relies on the Internet. The GeoNet is shown in the upper plane.

The STP, an abstract specification, should not couple to the mode of communication (that is, is whether the service agents using the STP are communicating in M2M mode or Internet-based mode). The network of service agents encompassing both Internet-based and M2M links, operating on the basis of STP is called the *Service Web* and is illustrated in Figure 10.6. The Internet-based communication links are shown as solid lines and the M2M communication links are shown as dashed lines.

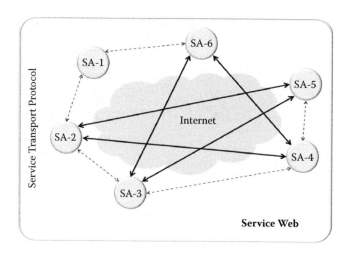

FIGURE 10.6
An illustration of a Service Web. The circles indicate service agents, the solid lines Internet-based communication links and the dashed lines M2M links. We have assumed that SA-1 does not have direct access to the Internet, while all the other nodes do.

I-2 is defined to be a globally connected network of interacting service agents that consume and provision web-enabled services using the Service Transport Protocol, subject to the following four constraints.

Constraint 1: The implementation details of the provisioned services and consumption details of the invoked services should be hidden behind service agents' uniform service interfaces, and should not be visible or accessible from outside the service agents.

Constraint 2: 1-hop communications between service agents should be of one of two types, as illustrated in Figure 10.6.

   a. **Internet-based** (heavy arrows). Ex. $SA_2 \rightarrow$ Internet $\rightarrow SA_5$.

   b. **M2M**[*] (dashed arrows). Ex. $SA_3 \rightarrow SA_2$. An M2M communication between two service agents, such as $SA_3 \rightarrow SA_2$, could rely on wireless channels such as air-based radio communication, or optical communication or even communication over a wired link. The only restriction on an M2M communication is that it should not use the Internet.

Constraint 3: Every service agent in I-2 should be connected to the Internet through a bidirectional path that could comprise M2M communication links. For example, SA_1, which is assumed not to have direct connectivity to the Internet, would still be regarded as belonging to I-2 since it is connected to the Internet through a path such as $SA_1 \rightarrow SA_2 \rightarrow$ Internet, where $SA_1 \rightarrow SA_2$ is an M2M link.

Constraint 4: Every service agent in I-2 should support the Service Transport Protocol (STP), and all interactions between service agents in I-2 are to be governed by the STP. In other words, interactions between service agents that are outside the scope of STP are to be regarded as out-of-band and forbidden within I-2.

Constraint 2 is a new feature that differentiates I-2 from the World Wide Web. Whereas the web relies wholly on the Internet for end-to-end connectivity, I-2 does not. Relaxing the restriction that connectivity must be Internet-based, by allowing the M2M communications, enables I-2 to fold the physical objects, which may not be IP-nodes, to be networked into the larger I-2 infrastructure.

Constraint 3 excludes island networks of service agents that lack bidirectional connectivity to the Internet from being included in I-2. I-2 is envisioned to be an integration of the cyber and physical worlds. Connectivity to the Internet, the scaffold of the cyber world, is hence an essential feature for a service agent to be included in I-2. Admittedly, embedding the necessary computing and communication capabilities inside service agents in

---

[*] Machine to machine.

order to satisfy Constraint 3 adds to the implementation costs. However, with advancing technology, we expect that the resources needed by service agents will become available at acceptable cost.

Armed with the above coarse-grained description of I-2, we compare the anatomy of I-2 with that of the Internet and the web in the next section. The comparison highlights the similarities among the three architectures.

The Internet, Web, and I-2

We present a coarse-grained comparison of the Internet, web and I-2 infrastructures, below. The three infrastructures bear considerable topological similarity to the *network of highways* connecting different *cities* in a country. Imagine that the only vehicles traveling on the highways are *trucks* that carry payloads among different cities. The network of highways is *payload-agnostic*, that is, the highways do not care what payloads are transported in the trucks. A truck on a highway is taken as the unit of transaction, in the *transportation infrastructure*.

Similarly, the *Internet* infrastructure is a network of communication highways connecting the *IP-enabled devices*. The payloads transported over these communication highways are the *IP datagrams*. The Internet does not care about the contents of the datagram any more than the network of highways cares about the contents of the trucks. That is, the Internet is also payload-agnostic. The transportation of IP datagrams on the Internet is governed by the TCP/IP transport protocol.

The *World Wide Web* is also a transportation network whose end nodes are called *components* in the REST style. Web servers and browsers are examples of components. The network supports the flow of *resources*. Again, the web infrastructure is payload-agnostic. It does not care about the semantics of the resources being transported. The transportation of resources is governed by HTTP, the HyperText Transfer Protocol. The semantics of the resources are deciphered by the applications running at the edge.

Finally, the I-2 infrastructure can also be viewed as a transportation network that transports *web-enabled services*. The end nodes of I-2 are *service agents*. Like the Internet and the web, the I-2 infrastructure just described is also *payload-agnostic*. It merely provides a substrate transportation infrastructure that enables the service agents to transport units of interaction, namely, web-enabled services. The I-2 architecture does not care about the details of the web-enabled services transacted over it. It places no selective barriers on any subgroup of web-enabled services, and is hence a *universal* infrastructure.

Umbrella Constructs and Architectural Simplicity

Figure 10.7 summarizes the analogy among the four infrastructures. Each of the four infrastructures achieves architectural simplicity by inventing umbrella constructs—*truck, IP datagram, resource*, and *web-enabled service*—that encapsulate and hide the heterogeneity from the core architecture of the infrastructures.

For example, the network of highways restricts itself to providing a platform for the transportation of trucks, leaving it to the cities and the end users therein to decide what gets put into the truck by the senders and how those contents are used by the receivers. The heterogeneity of contents, their uses, their semantics, are confined to the edge of the infrastructure. The core architecture of the network of highways is kept simple as just a platform for the transportation of trucks. Such architectural simplicity is achieved using an encapsulating umbrella construct called the truck.

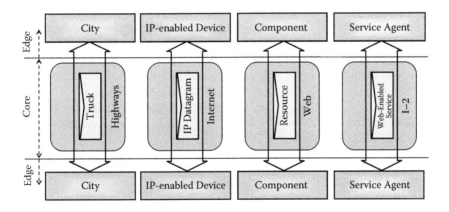

	Network of Highways	Internet	World Wide Web	Internet 2.0
Network node	City	IP-enabled Device	Client/Server	Service Agent
Unit of Transaction	Truck	IP Datagram	Resource	Web-Enabled Service
Transport Protocol	-	TCP/IP	HTTP	STP[2]

[2]Service Transport Protocol

FIGURE 10.7
Comparison of the Internet, web, and I-2.

The other three infrastructures also similarly achieve architectural simplicity at the core, thanks to the respective umbrella constructs. Specifically in the case of I-2, the details of the web-enabled services being transacted—such as whether their implementations involve physical objects, software resources, humans, or a combination of all three—are relevant to the service agents operating at the edge of the infrastructure. I-2's core, like the network of highways, is merely concerned with providing a substrate platform over which the interacting service agents can provision and consume web-enabled services. This separation of concern between I-2's architecture and the implementation of web-enabled services inside its end nodes—the service agents—is a critical feature in the design of I-2.

Anatomy of a Service Agent

I-2 places no restrictions on the functionalities of the service agents or the details about service provisioning and consumption within them. However, the service agents in I-2 must have the minimum set of features, described in this section. The anatomy of a service agent is shown in Figure 10.8.

The communication layer comprises the hardware and the implementation of the protocols needed for communication over the Internet and/or the GeoNet. The service interface layer houses the implementation of the Service Transport Protocol to facilitate the advertisement, description, discovery, and invocation of services. We will consider each of the activities separately in the following sections.

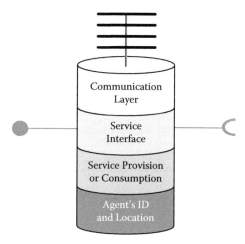

FIGURE 10.8
The structure of end nodes in Internet 2.0.

The service provisioning/consumption layer comprises the implementation of the provided web-enabled service—in service agents that provide services—and service consumption functionality—in service agents that consume web-enabled services. In hybrid service agents that provision as well as consume web-enabled services, the provisioning and consumption functionalities are housed in this layer. The bottom layer houses the information about the identity of the agent in the different networks in which it holds membership. It also houses the logic to determine the agent's current location, which would be static data if the agent is stationary or dynamic data, derived possibly through GPS, if the agent is mobile. The requirement that the agent carry the capability to determine its own current location data is a new feature that is peculiar to I-2 architecture.

As described above, I-2 can be viewed as a set of service agents whose interactions are governed by the *Service Transport Protocol* (STP). At present, STP does not exist, and one of the key tasks in designing I-2 is the formulation of the STP. In the following sections we outline a specification of the features that must be incorporated into STP.

Service Transport Protocol: Service Description

As with the web services, we make a distinction between discovery-oriented description and invocation-oriented description of a web-enabled service. Discovery-oriented description provides a semantic description of the service and keywords to facilitate its discovery. The invocation-oriented description, on the other hand, provides a syntactic description of the input/output interfaces of the service, intended to facilitate its invocation. We touch upon the salient features that must be incorporated into the STP to support both a discovery-oriented and invocation-oriented description of web-enabled services.

Discovery-Oriented Description: Semantic description of a web-enabled service is critical to its discovery by potential consumers. Semantic description could include keywords, verbal description, and classification codes of the service, derived from standard semantic taxonomies. The existing semantic taxonomies for services are too coarse-grained and not tailored to the universe of web-enabled services [UNSPSC 2012, NAICS 2012, and ISIC 2012]. One hopes that new and better taxonomies will emerge as the landscape of web-enabled services evolves.

The universe of web-enabled services being less heterogeneous than the space of general services, it is feasible to develop a customized *hierarchical semantic classification* scheme for web-enabled services, similar to the one described in Chapter 9 (in the context of web services). Such a classification scheme should be able to encompass most categories of web-enabled services. It will provide a powerful prefix-based classification code, facilitating search at various levels of granularity.

To accommodate the emergence of new web-enabled services and taxonomies without altering I-2 architecture, web-enabled services should be tagged with a *2-field classification code*: the classification scheme field and the service code within the classification scheme. For example, the pair (UNSPSC, 81111902) specifies the service class numbered 81111902 in the UNSCPSC classification scheme, that is, the online database information retrieval service [UNSPSC 2012].

Invocation-Oriented Description: The STP must also specify the *language* for invocation-oriented description of web-enabled services. The WSDL-based description that has been developed for web services provides a useful framework that can adapted for web-enabled services as well.

The service agents in I-2 will operate under a broad spectrum of resource constraints. A one-size-fits-all approach to designing the service protocol is bound to present high barriers for resource-constrained service agents. Therefore, STP's design must be *modular*, allowing resource-constrained service agents to implement a lighter reduced-function installation of the protocol while allowing the resource-rich service agents to avail of the full functionality.

Finally, the memory and power constraints on resource-constrained service agents could make it impractical for them to store and transmit the descriptions of their services. The STP must be designed to allow such service agents to store the service description files elsewhere on the I-2 network, and provide only the network address of the file in response to requests for service descriptions from potential service consumers.

Service Transport Protocol: Service Discovery

A requirement for a service agent to belong to I-2 is that it must have a bidirectional connectivity to the Internet. Accordingly, two service agents of I-2 would always be able to talk to each other using Internet-based communication channel. In addition, as shown in Figure 10.5, two service agents that are within each other's geographic proximity would also be able to communicate directly—for example, using direct wireless connection between them—without routing their messages through the Internet. Such opportunistic connectivity is indicated by dashed lines in the GeoNet plane in Figure 10.5. Opportunistic connectivity is a new feature of I-2 stemming from the mobility of the physical objects. In contrast, the Internet does not support opportunistic connectivity, requiring all communications between two service agents to flow through the Internet, even if the two IP-nodes are geographically close to each other.

In keeping with the two types of connectivity, in I-2 there would be two types of discovery mechanisms for services—*Internet-based discovery of services* and *opportunistic discovery of services*. The former category comprises services that are provisioned and consumed through Internet-based communication and the latter through opportunistic communication that does not involve the Internet.

Internet-Based Discovery of Services: Two models for Internet-based service discovery have been explored in the past—the *registry model* and the *search engine model*. In the registry model the provider registers the service with a central registry, such as the UDDI framework. A potential consumer then discovers a registered service by querying the registry. In the search engine model, a provider merely exposes a provisioned service on a server. A third-party search engine periodically scours the service web looking for new services and updates of existing services. The search engine maintains an index of provisioned services. A potential consumer can then discover the available services by querying the search engine's index.

The UDDI framework, which is based on the registry model, has not gained widespread support. Even its early supporters have decommissioned their UDDI service agents [UDDI 2006]. On the other hand, the search engine model has been wildly successful for the discovery of resources on the web. Admittedly, resources and services are different entities, and the extent of adoption of UDDI could be indicative of the peculiarities of the UDDI implementation and not of systemic shortcomings in the registry model. However, the registry model presents a higher barrier for expansion of the service web than the search engine model. Therefore, we recommend restricting attention to the search engine model in the design of STP.

Opportunistic Discovery of Services: This discovery mechanism assumes that the provider and consumer are geographically close to each other and can communicate, usually wirelessly, without having to route their communication through the Internet. The two different opportunistic discovery modes are the *provider-driven discovery* and *consumer-driven discovery.*

In *provider-driven discovery,* a provider service agent periodically broadcasts its presence and the web-enabled services it offers, enabling consumer service agents in its vicinity to discover its existence. In *consumer-driven discovery,* on the other hand, a consumer service agent broadcasts requests for the web-enabled service it needs at a given instant. If any service provider service agent, receiving the broadcast, offers the desired service then the provider service agent responds, advertising its service. In order for service providers and consumers to be able to listen for asynchronous broadcasts from each other they need to be running a common protocol. The STP suite must specify the protocol needed for both provider-driven and consumer-driven discovery.

Service Transport Protocol: Service Invocation

The web services provide a useful model for web-enabled services. Both are provisioned over electronic communication networks. Both provide services encapsulated behind standard interfaces. Their description, discovery, and invocation mechanisms also overlap enough to make the class of web services a useful template for web-enabled services. Beyond their similarities, however, the two families of services embody differences that are significant enough to necessitate a new protocol for the emerging web-enabled services. We begin the discussion with a critical review of the current trends in the invocation of web services that paves the way for a discussion of the features that must be incorporated within STP to support the invocation of the web-enabled services.

As we discussed in Chapter 9, the two prominent approaches for invoking web services are the resource-centric *RESTful approach* and process-centric *SOAP-based approach*. Both of the approaches rely on the main transport protocol for the web—the HTTP.

HTTP, the HyperText Transport Protocol, was designed for transporting hypertext documents, as the name suggests. It has been spectacularly successful in fulfilling the purpose for which it was designed. Of late, however, HTTP and the related notion of resource are being used for applications for which they were neither intended, nor are necessarily suited.

HTTP verbs are being used to invoke and provision the so-called RESTful web services, although many of those services have little or nothing to do with hypertext documents. The notion of resource, which was originally intended for hypertext documents on the web, is being stretched to cover digital, nondigital, and even abstract entities. The attempted enlargement of the scope of the term resource led to a contrived hash (#) notation for naming resources, and subsequently, to a contrived use of the HTTP status codes to distinguish between digital and nondigital resources [Berners-Lee 2009].

Even the RESTful web service paradigm is attempting to force-fit the notion of service into the notion of resource. The term "RESTful web service" also is being used for activities that are decidedly not RESTful. One of the principal features of REST, as discussed in Chapter 8, is that "hypermedia is to serve as the only engine of application state." RESTful web services are being consumed by applications that have little or nothing to do with hypermedia.

The argument for using HTTP-based invocation is that HTTP is already widely deployed. However, this argument is weakened by the forecast that in the coming years the number of devices that will connect to I-2 will greatly exceed the number of devices on which HTTP is deployed at present. While I-2 infrastructure is in its formative stages we have the opportunity to set it on a firm foundation by migrating to a protocol that is tailored to its needs, instead of recycling HTTP for an infrastructure for which HTTP was not designed.

The attempts to stretch the HTTP and the notion of resource beyond their originally intended uses signal the need for a protocol with enhanced functionality—namely, the *Service Transport Protocol*—and a new construct that has broader applicability than a resource—namely, *web-enabled service*. Continuing use of HTTP and the notion of resource to provision web-enabled services, ignoring the need for a new protocol, would actually hinder the development of I-2 infrastructure.

Against the backdrop of the invocation mechanisms for the web services, we list the features that must be incorporated into STP to support invocation of web-enabled services.

1. First, STP must provide *uniform interfaces* for all web-enabled services—whether the services encapsulate software functionalities, hardware functionalities, human components, or a combination thereof.

2. One of the features of HTTP-based RESTful web services that is appealing is its simplicity. A limited vocabulary of verbs in HTTP is used to invoke a wide variety of services. Such simplicity could be built into STP by providing a set of *verbs* (*à la* HTTP) that could encompass all the functionalities needed for invoking and provisioning web-enabled services.

3. STP must be an *abstract specification* and should not be tied to any particular mode of communication—wireless, wired, optical, or human-mediated—or any of the protocols used in such communications. It should not even be tied to the HTTP. It should operate at the *application layer of the OSI model* and should be able to interoperate with whatever choices the provider and consumer service agents use at the lower layers of the OSI model. The independence from the choices at the lower layers of the OSI model will endow the STP with the universality needed to support a diverse collection of service agents, while providing a uniform interface.

4. STP transactions must be *stateless.* That is, all messages exchanged during service invocation should be self-contained, containing all the information necessary for a provider service agent to perform the requested service. In other words, the service invocation occurring between two service agents should not rely on previous interactions between the service agents.

5. The design of STP should be modular permitting *graded deployment* in proportion to the needs of service agent. At the low end of the spectrum, the lightest version of STP must include the bare minimum set of functionalities. Services provisioned by many resource-constrained service agents such as, say, the temperature sensors, may not need the bells and whistles such as security and reliability. These resource-constrained service agents should be permitted to

deploy a reduced-function version of STP. At the other end, providers that require a larger set of features should be allowed to deploy full-function version of STP. STP should be designed to ensure that the common framework underlying its deployments enables interoperability among the service agents, regardless of their customized deployment of STP. The flexibility to deploy STP at different tiers with features and functionalities tailored to the needs and constraints of the service agents ensures that the service agents are not saddled with functionalities they do not need, and minimizes the overheads on resource-constrained service agents.

6. Finally, STP must be an *open protocol*. The initial protocol as well as the later enhancements must be made available as RFCs (Requests For Comments).

Closing the Loop

In Chapter 8, we discussed the key design principles of the Internet and the web for the purpose of incorporating those principles into I-2's architecture. To close the loop, we look at the features of I-2, discussed in this chapter, against the backdrop of the discussion in Chapter 8 to verify that I-2's architectural imperatives indeed embody the design principles gleaned from the Internet and the web.

I-2, as described, is a *universal* platform. It is not geared toward any particular type of web-enabled service, device, or application. In the spirit of *open-architecture networking* I-2 places no restrictions on the nature of web-enabled services provisioned, the platforms, technologies, or devices the service agents use, or the manner in which the web-enabled services are implemented and consumed within the service agents. The resolution scale chosen for I-2 embodies the principle of *substitutability* in that any service agent can be substituted by another that has the same functionality but different internal details without affecting the architecture. Adding new service agents to I-2 involves deploying the STP in the agent and ensuring connectivity to the Internet. The efforts and resources needed for expansion are proportional to the size of the expansion, showing that the architecture is *scalable*. The STP, the bedrock of I-2, is to operate at the application layer of the OSI model and is to be independent of the details in the lower layers. Hence, the protocol and the architecture of I-2 can be expected to *endure* even as the technology evolves in the lower layers.

As we discussed in a prior section the umbrella construct, namely, web-enabled service, confines all the heterogeneity in I-2 to its edge ensuring *simplicity* at the core of I-2's architecture. I-2's architecture thus becomes a simple transportation network among interacting service agents that transact web-enabled services.

STP, the bedrock protocol, is required to be *stateless*. Further, STP is mandated to be an *open protocol*. The STP is required to have a *modular* design permitting the deployment of only those features that are needed in a resource agent. The resulting flexibility *lowers the barrier for entry*, especially in resource-constrained service agents. Using the Internet as its bedrock ensures that the I-2 infrastructure inherits the *low barrier for expansion* built into the Internet.

Summary

The discussion in this chapter is the central message of this book. I-2 is a complex system. It is shown that choosing a resolution scale at which service agents emerge as the irreducible building blocks of I-2 leads to dramatic simplicity in I-2's core architecture while providing the flexibility necessary to integrate cyber and physical worlds. I-2 is thus viewed as a network of interacting service agents that transact web-enabled services, with all interactions governed by a universal Service Transport Protocol. All of the entropic complexity arising from the diversity of physical and nonphysical resources, and the heterogeneity of devices that connect to I-2, are thus pushed to the edge and encapsulated inside the service agents. The architectural imperatives presented in this chapter are also shown to embody the design principles underlying the Internet and the web.

11

Building a Prototype for I-2

One of the earliest events in the evolution of both the Internet and the web was the development of prototypes. The prototype of the Internet started as four nodes, interconnected by preexisting telephone lines [Leiner et al. 1997]. Similarly, the web began as a prototype that was built by Berners-Lee [Berners-Lee et al. 1996]. The histories of the Internet and the web, outlined in Chapters 3 and 4, show that the prototypes served as seeds that grew into the two global infrastructures.

In contrast, we do not yet have an infrastructure that can be regarded as a prototype of I-2. Absent a prototype, any attempt to architect I-2 would be an *open-loop* design process. That is, without a prototype we lack the means to vet the architectures that are being proposed for I-2. Building a prototype enables us to perform field tests and use results from the field tests to apply selection pressure on the different architectural alternatives. That is, a prototype would enable a *closed-loop* design of I-2's architecture. The development of a serious prototype for I-2 would therefore have to be one of the first milestones in the roadmap for building the infrastructure.

A prototype for I-2 is a mosaic that involves several interlocking pieces. In this chapter we will take a closer look at the different pieces. Just as the fledgling prototype for the Internet exploited the existing communication network infrastructure, such as the telephone lines, we show that one can also design a prototype of I-2 that exploits the preexisting infrastructures to reduce the construction costs.

The rest of the chapter is organized as follows. First, we discuss the salient components of a prototype of I-2. Next, we discuss the details of the emerging bridge between the physical world and the mobile devices. We also briefly describe the paradigm of Mobile Ad hoc NETworks (MANETs). Finally, we outline a blueprint for a prototype of I-2 that involves a marriage of the MANET paradigm and the emerging bridge between the physical world and the mobile devices.

Key Imperatives for the Prototype

In order to have a concrete backdrop for our discussion, we will use the example of a hypothetical cyber-physical system: a *smart library*. Consider a smart library in which the books are digitally enhanced with passive RFID tags. We will assume that the library offers the following service to its patrons. If a patron provides the details of the book at a terminal located in the library—called a *gateway*—then the gateway advises the patron whether or not the book is available in the library, and if it is available, where exactly it is located in the library at the current time. The latter service—physically locating an available book within the library—is important since the book may be shelved at a wrong location or may be lying anywhere in the library waiting to be shelved after use by another patron.

In order to determine whether the book is available in the library, the gateway invokes a database lookup service provided by a service agent, that we will call WSP (Web Service Provider agent), as shown in Figure 11.1. Given the details of a book, the WSP looks up the library's database and returns the status of the sought book to the gateway.

If the book is available, then the gateway broadcasts a message containing the book's tag ID to a network of service agents, denoted WeSP (Web-enabled Service Provider agent), that are geographically distributed throughout the library. We will assume that each WeSP houses an RFID reader and the service logic necessary to process the data stream from the reader. We will also assume that the network of WeSPs is deployed in a configuration that ensures that every point in the library is within the range of sufficiently many RFID readers to permit high-resolution computation of the physical location of every book in the library. The WeSPs are assumed to be stationary, with their location and identifier information housed in their electronic memory.

Upon receiving the tag ID of the book being sought each WeSP activates its reader, which in turn wirelessly interrogates its physical surroundings to

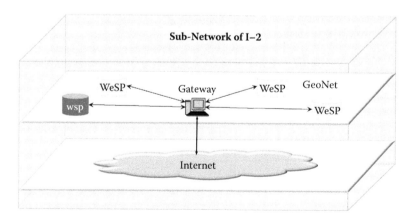

FIGURE 11.1
A subnetwork of I-2 illustrating the smart library infrastructure. WSP denotes Web Service Provider agent and WeSP Web-enabled Service Provider agent. The WSP does not involve interaction with the physical world, while WeSPs are capable of sensing physical objects.

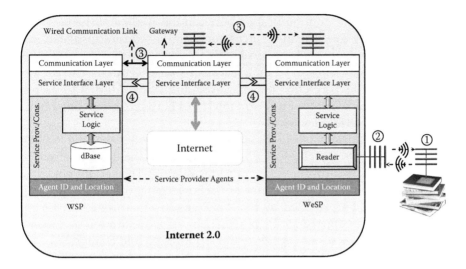

FIGURE 11.2
Finer structure of the smart library infrastructure.

detect the tagged books within its range. The tag IDs of all the books that are in the range of a reader are streamed to the service logic in the WeSP. The service logic then compares the tag ID sent by the gateway with the tag IDs detected by its reader and if there is a match reports discovery of the book to the gateway.

Figure 11.2, a combined representation of Figures 10.5 and 10.8, provides a fine-grained illustration of the I-2 subnetwork shown in Figure 11.1. Only one WeSP node is illustrated in Figure 11.2. Both WSP and WeSP have the four layers illustrated in Figure 10.8. The RFID reader and the service logic tied to it are housed in the Service Provision/Consumption layer of the WeSP. The WeSP provisions a single web-enabled service: given a tag ID, it returns a response, indicating whether or not the book with the given tag ID is in its vicinity. The STP-based uniform service interface of the WeSP is illustrated attached to its Service Interface layer. The WeSP communicates wirelessly with the gateway, while the WSP is shown connected to the gateway through a wired link. Both the WSP nodes and the WeSP nodes are capable of providing their identifiers and their locations, on demand.

The WSP and the WeSP are both service agents. However, they embody a difference: WeSP interacts with the physical world while WSP does not. Figure 11.2 also highlights an important aspect of I-2. The RFID-tagged books are not service agents and hence not the end nodes of I-2. They lack the resources necessary to run the Service Transport Protocol or to report information about their location. Rather the service agents, in the above example, are WSP and WeSP, both of which are equipped with all of the features that a service agent is required to have (see Figure 10.8). The books are physical objects, encapsulated behind the service agents.

The details of the communication between the WeSP and the tags—such as the frequency they use in their communication—are encapsulated in the service that a WeSP provides. Such details are pushed to the edge and are neither relevant nor visible to I-2 architecture. I-2 architecture is also designed to be blind to the distinction between WSP and WeSP. All that I-2 architecture sees are the STP-based uniform interfaces between interacting nodes at their Service Interface layers.

A second important aspect illustrated in Figure 11.2 is that the WeSP has a fluid coupling to the physical world. The books in the library are not confined to remain within the range of a single WeSP. The set of books that a WeSP couples to changes with time as the books move around in the library. The encapsulation provided by the WeSP ensures that the I-2 architecture itself remains unaffected as the physical objects move around and the coupling between I-2's service agents and the physical world changes.

In summary, there are five functionalities that we need to deploy in building a prototype of I-2. The functionalities are shown numbered in Figure 11.2.

1. The physical objects must be digitally enhanced to make them capable of communicating wirelessly with the service agents of the I-2 infrastructure.

2. We need to deploy two types of service agents: (1) service agents that can provide software-based services, such as the WSP in Figure 11.2, and (2) service agents that can interact with the physical world, such as the WeSPs. Whereas a WeSP serves mainly as a sensor agent, which funnels information from the physical world into I-2, the prototype must also include actuator agents that trigger action on physical objects.

3. We need to arrange for M2M-based, as well as Internet-based, connectivity among the agents; see the discussion in Chapter 10 for details.

4. Service Transfer Protocol must be developed and deployed in the service agents to create an infrastructure in which the agents have uniform interfaces and can interoperate.

5. Finally, and perhaps most importantly, the prototype must embody a low barrier, enabling even individuals with moderate to low technical skills to participate in driving the evolution of I-2 paradigm.

The fifth item listed above is intangible but is an important feature. Success of I-2 hinges as much on building the technical infrastructure as it does on the emergence of creative applications that harness the infrastructure. Web 2.0 shows that user-driven evolution of an infrastructure can dwarf the growth orchestrated by a smaller group of technical architects.

A Blueprint for a Prototype

In the smart library described above the service agents were deployed throughout the library. The theme park at Fort Lauderdale, discussed in Chapter 2, uses an I-2 infrastructure, similar to that of the smart library, to track visiting children. In the theme park, children, like books, are tagged with RFID wristbands. A network of I-2 service agents like WeSP, with embedded RFID readers, are stationed throughout the theme park to help locate children.

Deploying an infrastructure such as that in the hypothetical smart library or in the theme park on a global scale would be prohibitively expensive. And yet, Hewlett-Packard's Central Nervous System for Earth (CeNSE) initiative is attempting to do just that [Hewlett-Packard 2012]. The CeNSE initiative seeks to deploy about a trillion sensors and actuators all over the earth [Mone 2009]. The vision of CeNSE, or some variant of it, might one day become a reality. In the meanwhile, we outline below a less expensive approach to building a prototype for I-2, that harvests the unused duty cycles in the resources that are already deployed—the *wireless-capable mobile devices.*

Figure 1.2 shows that in 2010 there were about 78.6 mobile cellular subscriptions for every 100 people on the planet. At the end of 2012, the world population was about 7.056 billion [US Census 2012], and the number of cellular subscriptions had risen to about 6 billion [ITU 2012], representing an 85% penetration. Crudely, the statistics indicate that 8 out of every 10 people on the planet carry a cell phone. In the United States, 322 million cell phones are being used [CTIA 2012] by a population of about 314 million [U.S. Census 2012]. There is at least one cell phone in use for every person in the United States. As of June 2012, 41% (130.8 million) of the mobile devices in use within the United States are active smart phones or wireless-enabled PDAs; over 95% of the mobile devices (300.4 million) are active data-capable. Wireless-enabled tablets, laptops, and modems number about 21.6 million or about 6% of the mobile devices in use in the United States [CTIA 2012].

As impressive as the statistics inside the U.S. are, the most rapid growth in the cellular subscriptions is occurring outside the country. A sixth of the six billion active cellular subscriptions at the end of 2011, that is, about a billion subscriptions, are in India and China [ITU 2012].

These statistics suggest that the mobile cellular devices are in widespread use around the world. An increasing fraction of these cellular devices are smart phones or wireless-enabled PDAs; for example, within the United States the number of smart phones and wireless-enabled devices increased 37% from June 2011 to June 2012, while the number of wireless cellular subscriptions increased only 5% over the same duration [CTIA 2012]. In other words, the fraction of devices that are smart phones or wireless-enabled PDAs increased from 31% to 41% over a 12-month period.

Most of the cellular devices, particularly the smart phones and the wireless-enabled PDAs, come bundled with the four layers of functionalities required in the service agents of I-2. (See Figure 10.8.) For example, an increasing number of cellular devices that are in use are wireless-enabled, and are equipped to support both the Internet-based communications as well as the short-range M2M communications with nearby devices. Thus, they have the capabilities required in the *communication layer* of a service agent (see Figure 10.8). The devices also have sufficient on-board intelligence and memory. Thus, they can easily run a lightweight version of Service Transport Protocol and hence can provide *service interface* functionalities.

Modern cell phones have built-in digital cameras that can be used to *optically bridge* the cyber and physical worlds. The optical bridge is already being used in augmented reality applications in which, by pointing a camera at a physical object, one can retrieve semantic information about the target object (for example, see Wikitude [Perry 2008]). Some smart phones—for example, those made by Nokia—now come bundled with embedded *RFID readers* enabling smart phones to interact with the physical objects tagged with RFID labels [Roduner 2010]. NFC-capable phones provide yet another alternative to interacting with other NFC-capable devices as well as physical objects affixed with NFC-tags [NFC Forum 2012]. The wireless communication capabilities of smart phones are already being exploited to use them as versatile wireless remote controls, capable of operating multiple appliances such as televisions [Samsung 2012] and garage door openers [Liftmaster 2012].

Thus many of the modern cell phones come bundled with considerable amounts of memory and on-board intelligence as well as the built-in bridge technologies that enable them to connect with the physical objects. Hence, they are capable of supporting the functionalities needed at the *Service Provision/Consumption layer* of I-2 nodes.

Finally, cell phones have built-in *MAC* addresses that automatically give them unique identifiers. Modern cell phones also have built-in GPS receivers that enable the phones to determine their own locations and report their locations on demand. Thus, they have the capabilities needed in the *Agent ID and Location layer* of an I-2 service agent.

In summary, many of the modern cell phones have all of the functionalities needed at the four layers of an I-2 service agent (Figure 10.8). They are widely distributed with nearly 85% penetration into the world population. They are able to communicate with each other through the Internet, and also directly through M2M communications. The world population has already been trained to use the cellular phones. Therefore, using them as I-2's service agents lowers the barrier for user-adoption of I-2. They are currently not being used at 100% of their duty cycle providing an opportunity to harvest the available unused duty cycles. The devices are already operational for a different purpose, which means that there is little to no additional investment needed to use the devices for dual purposes. Hence, we propose the following roadmap for building the prototype.

> Build a prototype of I-2 using the fleet of currently operational mobile smart phones as the service agents of the infrastructure.

Description of the Blueprint

We suggest that the prototype be deployed in the context of university campuses. We may reasonably assume that many of the people operating on the campuses use mobile smart phones. The smart phones are wireless-capable, which, in principle, means that they can transmit messages to other smart phones within their range. In Figure 11.3 the smart phones are shown as

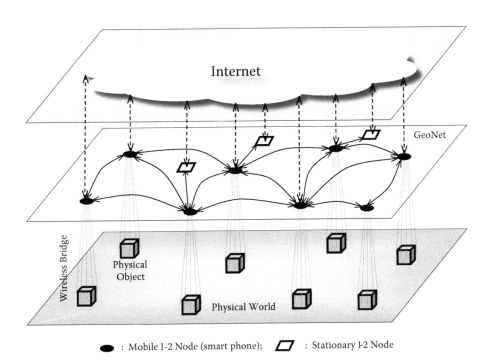

● : Mobile I-2 Node (smart phone);   ▱ : Stationary I-2 Node

FIGURE 11.3
Illustration of the proposed prototype of I-2.

oval service agents in the *GeoNet* layer. An arc between two service agents indicates that they are geographically close enough to exchange wireless messages. Since the smart phones are in general mobile, as they drift apart the arcs between them could disappear, even as new arcs emerge when the mobile devices approach each other. Thus, the GeoNetwork structure of the fleet of mobile phones on a campus could change dynamically as the people carrying these devices move around.

The parallelograms in the GeoNet plane represent stationary service agents distributed across the campus. An example of such stationary service agents is a wireless-capable library catalog server that enables mobile service agents to query the library's database wirelessly. When a mobile service agent and a stationary service agent are within each other's wireless range a communication link is established between the two agents. It is reasonable to assume that the mobile smart phones are connected to the Internet wirelessly, and the stationary service agents are also connected to the Internet through the university's gateways. The mobile and stationary service agents form the GeoNet layer of the prototype. We will assume that selected objects around the campus are digitally enhanced with RFID tags, sensors, actuators, and visual markers, enabling the service agents to sense and actuate them without human intervention.

The interesting aspect of the prototype is the GeoNet layer with dynamically changing topology. Although most of service agents in the GeoNet layer are connected to the Internet the connectivity through the GeoNet layer provides a capability—proximity-based M2M interaction—that is not easily realized through Internet-based connectivity. It is helpful to look at a few simple examples that demonstrate the value of the interagent connectivity in the GeoNet layer.

Consider an emergency scenario arising in some corner of a campus that could put the people throughout the campus at elevated risk. A message broadcast on the ad hoc network of mobile I-2 service agents would instantaneously reach nearly all the people on the campus. Broadcasting a similar message over the Internet is nearly impossible for obvious reasons. Even if the university were to maintain contact information for all the people affiliated with it and could broadcast a message to those mobile phones, it is difficult to reach the visitors on the campus. In such scenarios, the GeoNet is more useful than the Internet.

As a second example, consider the public transportation infrastructure in a university. If buses operating on campus are digitally enhanced with RFID tags, then the readers in the mobile smart phones could easily sense the nearby buses. The mobile phones that sense the location of a bus can also determine their own locations using their built-in GPS. Thus, the network of mobile phones could periodically broadcast information about the locations of buses across the entire ad hoc GeoNet, enabling people who use the public transportation to obtain accurate real-time information about the movements of the buses.

The previous examples involved only the mobile service agents. For an example involving mobile and stationary service agents, consider a parking garage on campus in which a stationary network of service agents is deployed. Each stationary service agent, let us assume, has an embedded RFID reader, is wireless-capable, is connected to the Internet, and has its location information housed in its software. A user with a mobile smart phone could wirelessly provide this network the tag ID of the RFID transponder in his/her vehicle and have the network return the location of the vehicle in the garage. Such a transaction involves an interaction between the stationary and mobile service agents. Alternatively, a user at a browser could invoke the same service over the Internet, which would involve interactions among only the stationary service agents.

The preceding examples demonstrate that the connectivity in the GeoNet plane can add functionality that is not easily realizable using only the Internet-based connectivity. The provision for service agents to communicate with each other over air-links, based on geographic proximity, is a key new distinguishing feature of I-2 that must be incorporated into the prototype.

In summary, the roadmap for building the prototype is to recruit the mobile smart phones that are already in use and make them function as the service agents of I-2. In order to use the mobile smart phones as service agents, we need to install two enhancements in them: (1) the Service Transport Protocol must be deployed in the participating phones, making them capable of functioning as providers/consumers of web-enabled services. (2) The phones must be enabled to support M2M communications over ad hoc networks that form in the GeoNet layer. The interagent communications in the GeoNet layer require the development and deployment of a new *GeoNet Communication Protocol (GCP)* designed specifically for such ad hoc networks.

The above two enhancements are essential to build the prototype. An optional third enhancement—the deployment of *middleware* on the phones—would promote user-driven growth of applications in the prototype's ecosystem. We elaborate on the need for middleware below.

Consider the two mobile smart phones shown in Figure 11.4. Assume that both of them contain the GPS system built into their hardware. Accessing the GPS resource in a phone requires knowledge of platform-specific details of the phone. On the other hand, if a middleware layer is installed in both the phones, exposing the GPS resource through a universal platform-independent interface, then the development of application software that invokes GPS would be greatly simplified. The application software would use the universal interface to the GPS in its code without having to worry about the platform-specific details involved in using the GPS.

In the previous chapter we discussed the features that must be incorporated into the Service Transport Protocol. In the remainder of this chapter we describe the features that must be incorporated into the GeoNet Communication Protocol. We begin by looking at the details of communication over the GeoNet. The interagent communications hinge on the

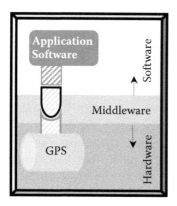 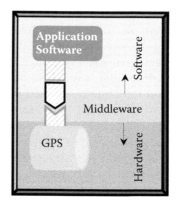

FIGURE 11.4
Middleware to expose hardware resources through universal interfaces.

addressing schemes. We discuss the addressing schemes, various forms of interagent communications in MANETs, and finally classify the MANETs, according to the mobility of its nodes. The discussions on the connectivity aspects of communication are finally encapsulated into a specification of requirements for the GCP.

We have restricted the discussion on GCP to those aspects that pertain to connectivity. The features of GCP pertaining to security and quality of service, although not discussed here, are important details that must be incorporated into the final design of the protocol.

Network Topology in GeoNets

We begin with a discussion of network topology in GeoNet. Consider five mobile phones within a geographic region, as shown in Figure 11.5. The range of a mobile phone—the maximum distance to which it can reliably transmit a wireless message (for example, over a Wi-Fi channel)—is shown as a circle centered at the phone. The phones could have different ranges, as indicated by circles of varying sizes. Since nodes A and B are within each other's range, they can engage in two-way communication. On the other hand, the communication between B and C is 1-way since B is outside C's range. Thus, the connectivity within the network is represented as a directed graph.

In Chapter 5 we discussed the wireless protocols, categorized by ranges. The WPAN (Wireless Personal Area Network) protocols, such as 6LoWPAN and ZigBee, are used in short-range communications (range ~10 meters). The WLAN (Wireless Local Area Network) protocols, such as Wi-Fi, are used in medium-range communications (range ~100 meters). The WWAN (Wireless

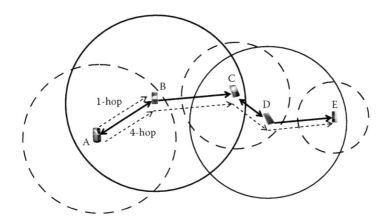

FIGURE 11.5
Ad hoc network of mobile smart phones.

Wide Area Network) protocols are used in long-range communications (range ~1 kilometer). The smart phones on the market support several protocols, and accordingly have different communication ranges.

The M2M communications in the GeoNet layer can be classified into two categories: *1-hop communication* and *multihop communication.* A direct communication from A to B is an example of a 1-hop communication, since the message does not go through any other intermediate phone. The figure also shows a 4-hop communication A $\rightarrow$ B $\rightarrow$ C $\rightarrow$ D $\rightarrow$ E. The message from A is routed by the intermediate phones B, C, and D to the final destination E. Such multihop communication is possible only if phones are enabled to serve as both the endpoints of communication—that is, senders/receivers—and also as intermediate routers.

An important feature of GeoNets is that the mobility of its nodes makes its network topology change. Figure 11.6 shows the impact of the movement of nodes C and E on the structure of the GeoNet. As a result of E's movement a message from A can now reach E in two hops, instead of the four hops needed in Figure 11.5.

The movement of node C disconnects the network shown in Figure 11.5 into two separate networks, as shown in Figure 11.6. In the GeoNet shown in Figure 11.6, nodes C and D cannot communicate with nodes A, B, and E. The connectivity between C and D, which was bidirectional in Figure 11.5, has also changed in Figure 11.6. While node D can still send messages to node C, it can no longer receive messages from C. The fluctuations in the topology of GeoNets resulting from the movement of its nodes makes routing messages in GeoNet a more challenging task than that in the Internet. Next, we turn to the different communication modes in GeoNets.

A multihop communication from a node in the GeoNet can be classified into two categories—*one-to-one* communication and *one-to-many* communication. See Figure 11.7. Each of these categories is further subdivided into

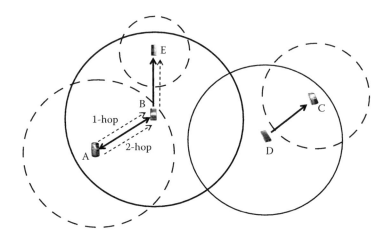

FIGURE 11.6
Impact of the movement of nodes C and E on the network structure. Compare with Figure 11.5.

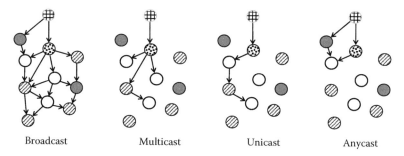

FIGURE 11.7
Various modes of communication.

two subcategories. One-to-one communications can be either *unicast* or *anycast*. In a unicast communication the message is sent to a specific destination node in the network. On the other hand, in an anycast communication the message is sent to one of a possible set of destination nodes. The one-to-many communication can be either a *broadcast* or a *multicast*. A broadcast communication is intended for all the nodes that receive the message.[*] It does not have specific target nodes. Broadcast is a popular communication mode when a node wishes to either advertise the service it offers, or wishes to advertise its request for a service. It is also useful when a node seeks to provision an unsolicited service by broadcasting useful information to the other nodes—such as information about some disaster in the geographic vicinity. A multicast on

---

[*] While it is a simple communication mode it could induce the emergent phenomenon of *broadcast storm*, which is an escalated contention for channels due to retransmissions of the original message. Several strategies have been proposed to avoid the broadcast storm effect, and make the broadcasting more efficient [Abdulai 2009].

the other hand is intended for several specific nodes, such as the nodes in a certain cluster. A special type of *multicast* is the *geocast* in which the set of target nodes lies within a specific geographic region. Figure 11.7 illustrates the four types of communications. In the multicast shown in the figure the target nodes have white interior. In the anycast communication shown the objective is to reach one of the nodes with white interior.

Node Addresses in GeoNets

Multihop communications, such as the unicast shown in Figure 11.7, require that the message be routed to a specific destination node in the GeoNet. In order to route a message to a specific destination node one needs an address for the destination node. We take a closer look at the issue of addressing nodes in a GeoNet.

Our interest is in using the smart phones as service agents of the I-2 prototype. So we will assume that the nodes of the GeoNet are smart phones. Although the smart phones come bundled with built-in GPS system, the geographic address of a phone is not helpful since it changes as the phone moves.

Instead, we could consider the virtual address assigned to a node when it joins a cluster in the GeoNet (see Chapter 5 for a discussion of cluster formation in ZigBee networks). The virtual address within a cluster could also vary with time.

For example, in Figure 11.5, assume that node A is the head of the cluster comprising nodes B, C, D, and E. Being members of A's cluster let us assume that the nodes are assigned virtual addresses (VA) as follows: VA(A) = 1, VA(B) = 2, VA(C) = 3, VA(D) = 4 and VA(E) = 5. Call the locations of nodes C and E in Figures 11.5 and 11.6, their initial and final positions, respectively. Assume that node C moves from its initial position to its final position before node E starts moving from its initial position. On account of C's movement, node E loses its connectivity to A even before it starts moving, and is dropped from A's cluster. A's cluster now has only two nodes, A and B, which retain their old virtual addresses. When E moves to its final location, shown in Figure 11.6, it rejoins A's cluster and could get assigned the first available address, say, VA(E) = 3. As a result of the movement of nodes C and E, not only does the network topology change, but even the virtual addresses assigned to nodes, such as E, in A's cluster could change. Hence, virtual addresses are not very stable either, and cannot serve as a reliable destination address in a unicast communication.

Several approaches have been suggested for address auto-configuration in MANETs. For example, the Prophet Address Allocation uses random numbers as addresses of the MANET nodes, generating random numbers using schemes that minimize the probability of generating repeated random numbers [Zhou et al. 2003]. The random number, assigned as an address to a mobile node, can be obtained by interrogating the node. For other auto-configuration networking schemes, see Cheshire and Steinberg [2006] and Oki et al. [2012].

Proactive, Reactive, and Hybrid Routing Protocols

GeoNets can be classified by the extent to which they maintain awareness of their own topology. At one end of the spectrum are *Proactive GeoNets* in which a set of one or more special nodes maintains awareness of the topology of the entire network. For example, the GeoNet could comprise clusters, each of which is managed separately by its cluster head. The cluster head would maintain information about the connectivity structure within the cluster, as well as the connectivity among cluster heads. Cluster heads are generally resource-rich nodes that can manage intracluster as well as intercluster communications. Information about the connectivity within a cluster as well as connectivity among clusters is periodically updated by the cluster heads by interrogating the network. Communications within such GeoNets is relatively straightforward. The routing protocols in which the information about the connectivity in MANETs is maintained proactively, even in the absence of communication activity among nodes, are called *proactive routing protocols* in the literature. Several proactive routing strategies have been proposed in the literature [Huang 2008].

At the other end of the spectrum are *Reactive GeoNets* in which none of the nodes maintains any awareness of the topology. This would be the situation when either all of the nodes lack the resources necessary to serve as cluster heads, or the mobility of the nodes is too high, making the frequent dynamic update of the topology information impractical. In such GeoNets, a route from a source node to a destination node is determined dynamically. For example, a path could be found by flooding the GeoNet with a broadcast from the source node. Each node receiving the broadcast registers the identity of the predecessor node from which it received the broadcast. When the destination node receives the broadcast it initiates a response that is routed back to the source node using the predecessor information stored by nodes on the path that led from the source to the destination. Such protocols, in which the route information is determined dynamically, as the need for a communication arises, are called *reactive routing protocols* [Huang 2008].

Finally, a *Hybrid GeoNet* operates between the two extremes mentioned above, maintaining some but not all of the connectivity information. The protocols for routing in hybrid GeoNets are called *hybrid routing protocols* [Huang 2008]. For example, in the hybrid protocol called the *Zone Routing Protocol*

(ZRP), the GeoNet is decomposed into zones [Huang 2008]. Connectivity information is maintained proactively within zones, but not across zones. Routing within each zone is done proactively while routing between zones is handled by a reactive routing protocol.

GeoNets can be categorized by the extent of mobility of nodes within the network. In a *quasi-static GeoNet,* typically the nodes move slowly and the topology changes occur over long time periods. In such GeoNets it is advantageous to maintain the network connectivity information proactively. On the other hand, a *dynamic GeoNet* is characterized by rapid movements of nodes, and rapid fluctuations in the network topology. The costs of frequent updates being too high, reactive protocols would be better suited for dynamic GeoNets.

In the envisioned I-2 prototype the plan is to harvest the unused duty cycles in smart phones on university campuses to build a dynamic distributed I-2 infrastructure. In such a prototype, the fleet of smart phones act as a distributed network of contact points between the cyber and physical worlds. In addition to the Internet-based connectivity, which comes prepackaged into the smart phones, the phones are also expected to support M2M connectivity in the GeoNet plane. The GeoNet Communication Protocol (GCP), described in this section, would govern the M2M communications in the GeoNet plane.

GCP for GeoNet can be viewed as the analog of TCP/IP for the Internet. Like TCP/IP, the GCP will also be based on the packet switching paradigm, and as such elements of TCP/IP can be adapted for use in GCP. However, GCP operates in a far more challenging environment than TCP/IP. In the remainder of this section, we summarize the preceding discussion to provide a noncomprehensive broad-brush specification of the essential features that must be built into GCP.

1. GCP must provide data packaging guidelines for GeoNet communications.

2. GCP must provide the functionalities to support different modes of data transmission (one-to-one and one-to-many).

3. GCP must be designed to interoperate with the prominent wireless protocols (such as the WPAN, WLAN, and WWAN protocols).

4. In a multihop communication, consecutive links may be based on different wireless protocols. The wireless protocols supported by a phone is a machine-dependent feature. GCP must have the capability to switch between wireless protocols in consecutive hops of a multihop communication.

5. At the data link layer GCP must contain the protocol for ensuring reliability of 1-hop transmission.

6. The GCP must include the functionalities necessary to enable a node to participate in the recurring network-wide attempts to gather the information about the connectivity structure of the GeoNet.

7. GCP must be designed to support any routing algorithm. Routing strategies, especially in a dynamic environment such as in GeoNets, can give rise to cycling. The GCP must embody features to detect cycling of packets.

8. GCP must be designed to support any address auto-configuration scheme.

9. In addition to reliability in 1-hop communications, GCP must also embody the functionalities to ensure end-to-end reliability in multi-hop communications. This is a significantly more challenging task in GeoNets owing to the node movements and the consequent changes to the network topology.

10. GCP must have the capability to handle the node's concurrent memberships in multiple clusters/networks.

11. The envisioned prototype seeks to scavenge the unused duty cycles in smart phones. GCP must provide the participating phones the option of limiting the amount of GeoNet data traffic that can flow through it.

12. Finally, and most importantly, the GCP must provide guarantees about the security and privacy of the participating phones. Understandably, the success of the prototype hinges on the credibility of such guarantees.

Summary

Building a prototype for I-2 is a critical first step in midwifing the birth of the infrastructure. In this chapter, we have presented a blueprint for a prototype that seeks to harvest the unused duty cycles in the smart phones operating on university campuses. The fleet of operational smart phones on university campuses provides a large network of geographically distributed connection points between the cyber and physical worlds. Further, their ongoing pervasive use makes them ideal vehicles for propagating the I-2 technology. Building an I-2 prototype using the fleet of mobile smart phones involves augmenting the Internet-based communication capabilities of the smart phones with the capabilities for proximity-based wireless communications. We have presented an overview of the new communication protocol that needs to be developed and deployed to forge a *GeoNet*, a network in which internode communication relies on air-based wireless links between nearby mobile devices.

12

The Road Ahead

As mentioned in the preface, the discussion in this book revolves around three questions. (1) *What are the technical roadblocks for the emergence of I-2?* (2) *What are the features that must be incorporated into I-2's architecture?* (3) *How does one build a prototype of I-2?* In the quest for answers to these questions we have reviewed the design and evolution of two successful global infrastructures—the Internet and the web. The lessons embodied in the Internet and the web, as well as the discussion in the preceding chapters, are consolidated into the following set of recommendations that is intended to facilitate the ongoing efforts to build I-2.

Incubation Environment

The incubation of the I-2 infrastructure must be restricted to occur in an academic research environment, until its development reaches the commercialization phase.

The participation of commercial organizations is critical for sustaining a global infrastructure beyond the commercialization phase. (See Figure 7.3.) However, the lessons embodied in the histories of the Internet and the web suggest that in the incubation phase the design activity must be allowed to focus, unfettered, on the technical aspects and sheltered from possible stultification by competing business interests. The Internet and the web were successfully incubated within academic research environment. Restricting the incubation of the I-2 infrastructure to occur in an academic environment will streamline its birthing process as well.

Prototype

One of the first milestones, if not the very first milestone, in building the I-2 infrastructure must be the construction of a sizable general-purpose open prototype.

As discussed in Chapter 11 a prototype serves two critical functions. It provides a test bed for evaluating design alternatives. Absent a prototype, development of I-2 would be based on an open-loop design process. Second, the field experience with a sizable prototype will help identify the features that must be incorporated into the design to make it reliable and scalable.

In Chapter 11, we also discussed a possible blueprint for a prototype. Building the prototype involves the development of the GeoNet Communication Protocol. The blueprint discussed in Chapter 11 exploits the existing infrastructure, and thereby lowers both the economic barrier as well as the barrier for adoption of the prototype.

Architecture

The irreducible units of I-2's architecture, that is, its end-nodes, must be service agents. The units of transaction in the architecture must be web-enabled services.

Containing the explosion of entropy stemming from the heterogeneity of physical and nonphysical resources is critical for I-2's architecture. The service agent construct provides the abstraction necessary to hide the heterogeneity at the edge. Web-enabled services provide the abstraction that can subsume the diversity of interactions among the nodes.

The Service Transport Protocol, the bedrock of I-2's architecture, must be developed using the prototype as the test bed.

The Service Transport Protocol (STP) comprises components related to description, discovery and invocation of web-enabled services, as described in Chapter 10.

A hierarchical semantic classification scheme for web-enabled services must be developed to facilitate the discovery of services.

Stewardship

The ongoing endeavors to build I-2 must be defragmented, and the task of building a prototype must be entrusted to a community of academic researchers, backed by requisite federal funding.

I-2 bears a greater similarity to the Internet than it does to the web in that it requires deployment of new hardware in addition to the development of protocols and software. Therefore, federal funding is critical to the development and a broader adoption of the prototype.

> The federal agencies must remain engaged in a partnership with the academic community, to steward the development of the infrastructure until it reaches the commercialization phase.

Crowdsourcing the Evolution

> It is critical that the barriers for using the I-2 infrastructure and for developing applications on it should be kept low from the very beginning, starting with the prototype.

Lowering the barrier for using I-2 involves developing intuitive, user-friendly interfaces to the infrastructure. Lowering the barrier for applications development involves developing the middleware for the physical and nonphysical components of I-2, and user-friendly applications development environments.

Crowdsourcing the evolution of the infrastructure has two benefits. It enlarges the user base, making the infrastructure more enticing to commercial investors. Lowering the barrier for applications development taps into the creative energy of a larger community and will trigger the critical *second wave of applications.*

Epilogue

Internet 2.0 seeks to connect the physical world to the information processing power in the cyber world. It is reasonable to expect that the behavior of such an ecosystem cannot be predicted *a priori.* Smart objects that make independent decisions, when deployed to cooperate and often compete with similarly enabled smart objects, will give rise to unanticipated collective behavior. While some of the simple cooperative behaviors, such as the interactions between the thermostat and the furnace in a house, will enhance the living conditions, yet other collective behaviors, such as the congestive collapse of the Internet (discussed in Chapter 1), could have far-reaching detrimental effects. The appearance of unanticipated behavior in the interactions of a large number of components in a complex system falls in the purview of a fascinating and poorly understood discipline called *emergence*.

As we engage in the endeavor to build the large complex system—the I-2 infrastructure—it is altogether appropriate to be mindful of the sobering possibility that the complex system could, and very likely would, display emergent behavior that we do not foresee. Refraining from speculations about the possible emergent behaviors that we might see in I-2, we present, instead, a few simple examples of emergent behaviors in the following paragraphs. These examples illustrate the limitations of the **constructionist** approach—that is, the attempt to predict the behavior of a complex system based on an understanding of its constituent parts.

A pervasive, and a simple, example of emergent behavior is the capability for **self-replication** in living systems. An atom cannot replicate itself. Neither can simple molecules, or individual organelles in a cell. And yet, when a collection of atoms, molecules, and organelles come together to form a single-celled microbe, the composite system displays a magical new capability to make a copy of itself. The self-replication capability thus emerges at the level of a cell, and is irreducible to the properties of its constituents in the sense that it cannot be gleaned from an understanding of the smaller structures that make up the cell.* The science of emergence is the study of properties, such as self-replication, that abruptly emerge at a certain level of organization and complexity, but cannot be found in lower levels of organization, and cannot be reduced to the properties of the constituents.

Another simple example of emergence of hierarchical properties is provided by the interaction of vehicles in a traffic. Consider an instance of four vehicles traveling in adjacent lanes as shown on the left in Figure 12.1. Although the traffic in adjacent lanes constrains each vehicle to remain in its lane, the vehicles in adjacent lanes do not constrain the motion within the

FIGURE 12.1
Schematic illustration of traffic jam and traffic gridlock.

---

* A virus, which is a simpler structure than a cell, can self-replicate inside a host. Even smaller structures, such as plasmids (replicons), are also known to have the capability to self-replicate within a host. We restrict attention to systems that can self-replicate without the help of other self-replicating systems (hosts).

vehicle's lane. Thus, each vehicle is allowed to travel at its own speed and under such circumstances no vehicle experiences a traffic jam.

In a separate scenario in which all the vehicles are confined to a single lane, as shown in the middle in Figure 12.1, the speed of the vehicle in the front could constrain the speeds of all the vehicles behind it. If the vehicle at the front is moving more slowly than the vehicles behind it, then one observes the emergence of a traffic jam. It is a phenomenon one cannot observe when there is only one vehicle per lane. Rather it arises from the entanglement of the dynamics of vehicles.

Next, consider the gridlock illustrated on the right in Figure 12.1. If traffic flows in the directions indicated, on the roads around a block, then not only can the four lanes give rise to traffic jams, but the traffic jams themselves could get entangled leading to the emergence of a new phenomenon—the *gridlock*. In the gridlock shown, each of the four traffic jams is waiting for the others to clear. That is the traffic jams have entered a *deadlock*, waiting for an event that cannot happen.

A gridlock is a fundamentally different phenomenon than a traffic jam. While the building blocks of a traffic jam are interacting vehicles, the building blocks of a gridlock are interacting traffic jams. The hierarchy of emergent phenomena is illustrated in Figure 12.2.

Whereas a typical gridlock involves relatively few interacting components, the emerging I-2 infrastructure is expected to have billions of interacting service agents, many of them capable of making autonomous decisions. The entanglement of the behaviors of the autonomous service agents of I-2 can be expected to give rise to emergent properties of immense richness and complexity the foreknowledge of which cannot be gleaned from an

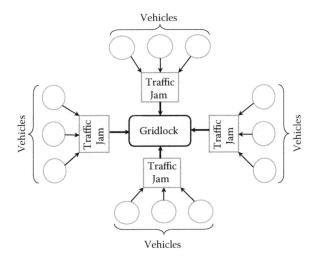

FIGURE 12.2
Hierarchical emergence of new phenomena at successive levels of complexity.

understanding of the behaviors of the individual components. Therefore, as the I-2 infrastructure evolves and comes of age, the reductionist focus on the characteristics of the individual pieces in the I-2 mosaic must give way to focus on an understanding of its possible emergent behaviors, on the design principles that eliminate undesirable emergent behaviors and induce the desired emergent behaviors. What surprises I-2 has in store, we do not know. As Niels Bohr said, "Prediction is very difficult, especially about the future."

References

Aaland, M., and Burger, R. (1992), *Digital Photography*, Random House, NY.

Abdulai, J. (2009), *Probabilistic Route Discovery for Wireless Mobile Ad Hoc Networks (MANETs)*, Ph.D. thesis, Information and Mathematical Sciences, University of Glasgow.

ABI (2010), *RFID Market to Reach $5.35 Billion This Year*, http://www.abiresearch.com/press/1618-RFID+Market+to+Reach+$5.35+Billion+This+Year (accessed July 30, 2012).

Akkiraju, R., Farrell, J., Miller, J., Nagarajan, M., Schmidt, M., Sheth, A., and Verma, K., *Web Service Semantics—WSDL-S* (2005), http://www.w3.org/Submission/WSDL-S/#Intro (accessed on November 10, 2012).

Akyildiz, I. F., Pompili, D., and Melodia, T. (2005), Underwater acoustic sensor networks: Research challenges, *Ad Hoc Networks* (3), pp. 257–279.

AlertMe (2012), www.alertme.com (accessed on September 20, 2012).

Alonso, G., Casati, F., Kuno, H., and Machiraju, V. (2010), *Web Services: Concepts, Architectures and Applications (Data-Centric Systems and Applications)*, Springer.

Alpert, J., and Hajaj, N. (2008), *We Knew the Web Was Big…*, http://googleblog.blogspot.com/2008/07/we-knew-web-was-big.html (accessed on December 18, 2011).

Arduino (2012), www.arduino.cc (accessed on October 14, 2012).

Arrayent (2012), *Arrayent Internet-Connect Service Overview*, http://www.arrayent.com/internet-connect.php (accessed on October 14, 2012).

Ashton, K. (2009), *That 'Internet of Things' Thing*, http://www.rfidjournal.com/article/view/4986 (accessed on September 18, 2012).

Ashton, K. (2011), *Whither the Five-Cent Tag?*, http://www.rfidjournal.com/article/view/8212 (accessed on October 17, 2012).

Baker, M., and Nottingham, M. (2004), *The Application/Soap+Xml Media Type*, http://tools.ietf.org/pdf/rfc3902.pdf (accessed on September 12, 2012).

Balkesen, C. (2008), *EPC Network—An Internet of Things Infrastructure, in Business Aspects of the Internet of Things*, Seminar of Advanced Topics, FS 2008, Florian Michahelles, http://www.inf.ethz.ch/personal/cagri.balkesen/pdf/iot_seminar_2008_proceedings.pdf (accessed on October 10, 2012).

Banks, J., Pachano, M., Thompson, L., and Hanny, D. (2007), *RFID Applied*, John Wiley & Sons, Hoboken, NJ.

Barish, G., and Obraczke, K. (2000), World Wide Web caching: Trends and techniques, *Communications Magazine IEEE*, Vol. 38, Issue 5, pp. 178–184.

Berners-Lee, T. (2009), *A Short History of Resource in Web Architecture*, http://www.w3.org/DesignIssues/TermResource.html, August 2, 2009 (accessed December 12, 2011).

Berners-Lee, T. (2004), *New Top Level Domains.mobi and .xxx Considered Harmful*, http://www.w3.org/DesignIssues/TLD (accessed on September 25, 2012).

Berners-Lee, T. et al. (1998), *Uniform Resource Identifiers (URI): Generic Syntax*, RFC 2396, http://www.ietf.org/rfc/rfc2396.txt (accessed on March 27, 2013).

Berners-Lee, T. (1996), *The World Wide Web: Past, Present and Future*, http://www.w3.org/People/Berners-Lee/1996/ppf.html (accessed on Feb 18, 2013).

Berners-Lee, T. et al. (1996), *Hypertext Transfer Protocol—HTTP 1.0,* RFC 1945, http://tools.ietf.org/html/rfc1945 (accessed on October 28, 2012).

Berners-Lee, T. (1989), *Information Management: A Proposal,* http://www.w3.org/History/1989/proposal.html (accessed on February 17, 2013).

Bhushan, A. (1971), *File Transfer Protocol,* http://tools.ietf.org/html/rfc114 (accessed on July 14, 2013).

BlueTooth (2012), http://www.bluetooth.com/Pages/Basics.aspx (accessed on September 23, 2012).

Botanicalls (2012), *The Plants > The Brains: The Sensors and Network > The Phone,* http://www.botanicalls.com/classic/how-it-works/ (accessed on September 19, 2012).

Burnham, M. (2009), *NASA-Cisco Climate Project to Flash 'Planetary Skin',* http://www.nytimes.com/gwire/2009/03/03/03greenwire-nasacisco-project-to-flash-planetary-skin-9959.html (accessed on October 11, 2012).

Buschmann, F. et al. (1996), *Pattern-Oriented Software Architecture Volume 1: A System of Patterns,* Wiley.

CASAGRAS (2009), *Final Report: RFID and the Inclusive Model for the Internet of Things,* CASAGRAS an EU Framework 7 Project, http://www.grifs-project.eu/data/File/CASAGRAS%20FinalReport%20(2).pdf (accessed on October 15, 2012).

Casaleggio Associati (2011), *The Evolution of Internet of Things,* http://www.casaleggio.it/pubblicazioni/Focus_internet_of_things_v1.81%20-%20eng.pdf (accessed on September 19, 2012).

CatchTheBusApp (2012), www.catchthebusapp.com (accessed on September 20, 2012).

Centos (2013), *Appendix C: Common Ports,* http://www.centos.org/docs/rhel-sg-en-3/ch-ports.html (accessed on March 4, 2013).

Cerf, V., and Kahn, R. (1974), A protocol for packet network intercommunication, *IEEE Transactions on Communications,* Vol. 22, No. 5, pp. 637–648.

Cerami, E. (2002), *Web Services Essentials (O'Reilly XML),* O'Reilly Media.

Cerf, V. (2010), *Greyglers@Google: Vint Cerf,* http://www.youtube.com/watch?v=t9M0RPNr9qg (accessed October 20, 2012).

Cheshire, S., and Steinberg, D. H. (2006), *Zero Configuration Networking: The Definitive Guide,* O'Reilly Media.

Chinnici, R., Moreau, J., Ryman, A., and Weerawarana, S. (2012), *Web Services Description Language (WSDL) Version 2.0 Part 1: Core Language,* http://www.w3.org/TR/wsdl20/ (accessed on September 12, 2012).

Cho, S. T., Najafi, K., and Wise, K. D. (1990), *Secondary Sensitivities and Stability of Ultrasensitive Silicon Pressure Sensors,* Tech. Digest. 1990 IEEE Solid-state Sensor and Actuator Workshop, Hilton Head, SC.

Cisco (2001), Cisco Mobile IP, http://www.cisco.com/warp/public/cc/pd/iosw/prodlit/mbxul_wp.pdf (accessed on September 24, 2012).

CISCO (2011a), *Cisco Visual Networking Index: Global Mobile Data Traffic Forecast Update, 2010–2015,* http://www.cisco.com/en/US/solutions/collateral/ns341/ns525/ns537/ns705/ns827/white_paper_c11-520862.pdf (accessed on December 20, 2011).

CISCO (2011b), *Entering the Zettabyte Era,* http://www.cisco.com/en/US/solutions/collateral/ns341/ns525/ns537/ns705/ns827/VNI_Hyperconnectivity_WP.pdf (accessed on December 20, 2011).

CNNMoney (2012), http://money.cnn.com/2012/10/04/technology/facebook-billion-users/index.html (accessed on October 16, 2012).

Collins, J. (2005), *RFID Delivers Newborn Security,* http://www.rfidjournal.com/article/view/1372 (accessed on August 9, 2012).

Comer, D. E. (2006), *The Internet Book: Everything You Need to Know about Computer Networking and How the Internet Works,* Addison-Wesley, Upper Saddle River, NJ.

Corella, M. A., and Castells, P. (2006), *A Heuristic Approach to Semantic Web Services Classification,* 10th International Conference on Knowledge-based and Intelligent Information and Engineering Systems (KES 2006), Springer Verlag Lecture Notes in Computer Science, Vol. 4253, Bournemouth, UK.

Corventis (2012), www.corventis.com; also see http://www.accessdata.fda.gov/cdrh_docs/pdf9/K091971.pdf (accessed on September 18, 2012).

CSI (2011), *Container Security Initiative, 2011,* http://www.cbp.gov/linkhandler/cgov/trade/cargo_security/csi/csi_brochure_2011.ctt/csi_brochure_2011.pdf (accessed July 29, 2012).

CTIA (2012), *Consumer Data Traffic Increased 104 Percent According to CTIA—The Wireless Association ® Semi-Annual Survey,* http://ctia.org/media/press/body.cfm/prid/2216 (accessed on December 4, 2012).

DART (2012), Deep-Ocean Assessment and Reporting of Tsunamis, http://nctr.pmel.noaa.gov/Dart/ (accessed on December 28, 2012).

DiYSE 2009, http://dyse.org:8080/display/hometest/About (accessed on October 14, 2012).

Dostalek, L. and Kabelova, A. (2006), *DNS in Action,* Packt Publishing, Birmingham, UK.

Dostalek, L., and Kabelova, A. (2006), *Understanding TCP/IP, A Clear and Comprehensive Guide to TCP/IP Protocols,* Packt Publishing.

Doukas, C. (2012), *Building Internet of Things with the Arduino (Volume 1),* CreateSpace Independent Publishing Platform.

Droms, R (1997), *Dynamic Host Configuration Protocol,* RFC 2131, http://tools.ietf.org/html/rfc2131#section-7 (accessed on September 13, 2013).

Dumas, M., O'Sullivan, J., Heravizadeh, M., Edmond, D., and ter Hofstede, A. (2003), *Towards a Semantic Framework for Service Description,* in Semantic Issues in E-commerce Systems: Ninth Working Conference on Database Semantics, April 25–28, 2001, Hong Kong, IFIP Conference Proceedings, ed. Meersman, R., Aberer, K., and Dillon, T. S., Vol. 239, pp. 277–291, Kluwer.

Dver, A. (2007), *Beaches and Theme Parks and Fairs—Oh My!,* http://missingchildprevention.files.wordpress.com/2007/03/beaches-and-theme-parks-and-fairs-oh-my.pdf (accessed on July 29, 2012).

Eckert, M., Bry, F., Brodt, S., Poppe, O., and Hausmann, S. (2011), A CEP Babelfish: Languages for complex event processing and querying surveyed, reasoning in event-based distributed systems; *Studies in Computational Intelligence,* Vol. 347, pp. 47–70.

Edwards, J., and Bramante, R. (2009), *Networking Self-Teaching Guide, OSI, TCP/IP, LANs, MANs, WANs, Implementation, Management and Maintenance,* Wiley Publishing.

Elevator World (2012), *The Elevator World, Timeline,* http://www.theelevatormuseum.org/timeline.php (accessed on December 25, 2012).

Engelbart, D. (2000), *National Medal of Technology,* http://dougengelbart.org/honors/nmt.html (accessed on March 4, 2013).

EOL (2012), *Energy Optimizers Limited,* http://www.superscout.com/Energy-Optimizers-Limited (accessed on September 19, 2012).

EPoSS (2013), *http://www.smart-systems-integration.org/public* (accessed on July 18, 2013).

Ergen, S. C. (2004), *ZigBee/IEEE 802.15.4 Summary*, http://pages.cs.wisc.edu/~suman/courses/838/papers/zigbee.pdf (accessed on September 24, 2012).

Evrythng (2012), http://evrythng.com/about (accessed on October 15, 2012).

FCC (2002), *Report of the Unlicensed Devices and Experimental Licenses Working Group*, http://transition.fcc.gov/sptf/files/E&UWGFinalReport.pdf (accessed on January 15, 2012).

Ferris, J. (2010), *Ideas That Changed the World*, DK Publishing, NY.

Fielding, R. T. (2000), *Architecture Styles and the Design of Network-Based Software Architectures*, Ph.D. Dissertation, University of California, Irvine, http://www.ics.uci.edu/~fielding/pubs/dissertation/top.htm (accessed on October 28, 2012).

Fielding, R. T., and Taylor, R. N. (2002), Principled design of the modern web architecture, *ACM Transactions on Internet Technology*, Vol. 2, No. 2, pp. 115–150.

Fielding, R. T., et al. (1999), *Hypertext Transfer Protocol—HTTP 1.1*, RFC 2616, http://tools.ietf.org/html/rfc2616 (accessed on October 28, 2012).

Fielding, R. et al. (1999), *RFC 2616: Hypertext Transfer Protocol—HTTP/1.1*, http://tools.ietf.org/html/rfc2396 (accessed on October 28, 2012).

Finkenzeller, K. (2010), *RFID Handbook: Fundamentals and Applications in Contactless Smart Cards, Radio Frequency Identification and Near-Field Communication*, 3rd Edition, John Wiley & Sons.

Floyd, S. (2000), *RFC-2914: Congestion Control Principles*, http://tools.ietf.org/html/rfc2914 (accessed on December 21, 2011).

Flynn, S. E., and Kirkpatrick, J. J. (2006), *The Limitations of the Current U.S. Government Efforts to Secure the Global Supply Chain against Terrorists Smuggling a WMD and a Proposed Way Forward*, http://www.cfr.org/border-and-ports/limitations-current-us-government-efforts-secure-global-supply-chain-against-terrorists-smuggling-wmd-proposed-way-forward/p10277 (accessed on July 29, 2012).

FP7 (2012), *What Is Fp7? The Basics*, http://ec.europa.eu/research/fp7/understanding/fp7inbrief/what-is_en.html (accessed on September 30, 2012).

Fuller, V., Li, T., Yu, J., and Varadhan, K. (1993), *Classless Inter-Domain Routing (CIDR): An Address Assignment and Aggregation Strategy*, http://www.hjp.at/doc/rfc/rfc1519.html (accessed on March 27, 2013).

Gambon, J., *RFID Frees Up Patient Beds*, http://www.rfidjournal.com/article/view/2549 (accessed on August 7, 2012).

Garfinkel, S., and Holtzman, H. (2006), Understanding RFID technology, in *RFID Applications, Security and Privacy*, eds: Garfinkel, S., and Rosenberg, B., Addison Wesley.

Garfinkel, S., and Rosenberg, B. (2006), *RFID Applications, Security and Privacy*, Addison Wesley.

Garlan, D., and Shaw, M. (1994), *An Introduction to Software Architecture*, http://www.cs.cmu.edu/afs/cs/project/able/ftp/intro_softarch/intro_softarch.pdf (accessed on October 2, 2012).

GEO (2012), http://www.earthobservations.org/about_geo.shtml (accessed on October 10, 2012).

GEOSS (2012), http://www.earthobservations.org/geoss.shtml (accessed on October 11, 2012).

Gerry Weber (2012), *RFID at Gerry Weber*, http://www.gerryweber.com/ag-website/en/ag-website/company/company-profile/rfid (accessed on August 6, 2012).

Gralla, P. (2006), *How the Internet Works?* Que Publishing.

Guinard, D., Ion, I., and Mayer, S. (2011), *In Search of an Internet of Things Service Architecture: REST or WS-*? A Developer's Perspective,* Proceedings of Mobiquitous 2011 (8th International ICST Conference on Mobile and Ubiquitous Systems), pp. 326–337, Copenhagen, Denmark, December 2011.

Guinard, D., Trifa, V., Karnouskos, S., Spiess, P., and Savio, D. (2010), Interacting with SOA-based Internet of things: Discovery, query, selection, and On-demand provisioning of web services, *IEEE Transactions on Services Computing,* Vol. 3, No. 3, pp. 223–235.

Hewlett-Packard (2012), *Intelligent Infrastructure Laboratory,* http://www.hpl.hp.com/research/intelligent_infrastructure/ (accessed on December 4, 2012).

Hill, M. D. (1990), What is scalability?, *ACM SIGARCH Computer Architecture News,* Vol. 18, Issue 4, pp. 18–21.

Huang, H. (2008), *Efficient Routing in Wireless Ad Hoc Networks,* Ph.D. Dissertation, Department of Computer Science, University of Arizona.

Hye-Bong, C. (1993), Typography in Korea, *Koreana,* Vol. 4, No. 2.

IBM (2004), *Item-Level RFID Technology Redefines Retail Operations with Real Time Collaborative Capabilities,* ftp://service.boulder.ibm.com/software/emea/dk/frontlines/RFID_IBM.pdf (accessed on August 6, 2012).

IBM (2007), *IBM Solution for Pharmaceutical Track and Trace,* http://www.alientechnology.com/docs/WP_IBM_Pharmaceutical.pdf (accessed on August 7, 2012).

IBM (2010), *IBM Delivers a New Level of Intelligence to the Expanding Mobile Business Market,* http://www.prnewswire.com/news-releases/ibm-delivers-a-new-level-of-intelligence-to-the-expanding-mobile-business-market-96473884.html (accessed on January 5, 2012).

IERC (2012), *IERC—The European Research Cluster on the Internet of Things,* http://www.internet-of-things-research.eu/about_ierc.htm (accessed on October 24, 2012).

IETF (2012), http://www.ietf.org/meeting/84/index.html (accessed on October 11, 2012).

Inoue, T., Hayakawa, A., and Kamei, T. (2011), *China's Initiative for the Internet of Things and Opportunities for Japanese Businesses,* NRI Paper No. 165, http://www.nri.co.jp/english/opinion/papers/2011/pdf/np2011165.pdf (accessed on October 11, 2012).

Internet Stats (2011), http://www.internetworldstats.com/stats.htm (accessed on October 17, 2012).

ioBridge (2012), http://www.iobridge.com/ (accessed on October 14, 2012).

IoT-A (2011), *Internet of Things Architecture, IoT-A Project Deliverable D1.1—SOTA Report on Existing Integration Frameworks/Architectures for WSN, RFID and Other Emerging IoT Related Technologies,* http://www.iot-a.eu/public/public-documents/documents-1 (accessed on March 27, 2013).

IoT-GSI (2011), *Internet of Things Global Standards Initiative (IoT-GSI),* http://www.itu.int/en/ITU-T/gsi/iot/Documents/tor-iot-gsi.pdf (accessed on February 13, 2012).

IPSO (2012), *IPSO Alliance: Enabling the Internet of Things,* http://www.ipso-alliance.org/ (accessed on October 11, 2012).

Isenberg, D. S. (1999), The mother of all disruptions, *America's Network,* 103(11), 12.

ISIC (2012), *International Standard Industrial Classification of All Economic Activities Rev. 4*, http://unstats.un.org/unsd/cr/registry/regcst.asp?Cl=27 (accessed on November 9, 2012).

ISOC (2010), *The Internet Model and Ecosystem*, www.isoc.org/pubpolpillar/docs/fcc_20100217.pdf (accessed on December 21, 2011).

ITU (2009), *The World in 2009: ICT Facts and Figures*, http://www.itu.int/ITU-D/ict/material/Telecom09_flyer.pdf (accessed on February 17, 2012).

ITU (2012), *ITU Releases Latest Global Technology Development Figures*, http://www.itu.int/net/pressoffice/press_releases/2012/70.aspx (accessed on December 4, 2012).

Jacobson, V. (2009), *Congestion Avoidance and Control*, in Proceedings of SIGCOMM '88 (Stanford, CA, Aug. 1988), ACM.

Jiang, J.-R., and Yeh, M.-K. (2009), Anti-collision algorithm in RFID, in *RFID and Sensor Networks*, ed. Chen, J., CRC Press: Boca Raton, FL.

Kelly, G. J., and Latzko, E. (2006), *Thirty Years of Photosynthesis 1974–2004*, Springer.

Kleinrock, L. (1961), *Information Flow in Large Communications Networks*, RLE Quarterly Progress Report.

Kopparapu, C. (2012), *Load Balancing Servers, Firewalls and Caches*, John Wiley & Sons.

Korom, J. J. (2008), *The American Skyscraper, 1850–1940: A Celebration of Height*, Branden Books.

Leiner, B. M. et al. (1997), *Brief History of the Internet*, http://www.internetsociety.org/internet/internet-51/history-internet/brief-history-internet (accessed on January 23, 2012). Also see, *The Past and Future History of the Internet*, Communications of the ACM, vol. 40, no. 2, pp. 102–108, February 1997.

Lewis, F. L. (2004), Wireless sensor networks, in *Smart Environments: Technologies, Protocols and Applications*, ed. Cook, D. J. and Das, S. K., John Wiley & Sons, New York.

Liftmaster (2012), *The Best Garage Door Opener Just Got Better*, http://www.liftmaster.com/lmcv2/products/liftMasterMyQEnabledProducts.htm (accessed on December 4, 2012).

Lorenz, M., Muller, J., Schapranow, M. P., Zeier, A., and Plattner, H. (2011), Discovery services in the EPC network, in *Designing and Deploying RFID Applications*, ed. Cristina Turcu, ISBN 978-953-307-265-4, InTech.

Mainz (1997), *The Gutenberg Mainz*, http://www.gutenberg.de/g2000.htm (accessed on October 2, 2012).

Mandel, L. (2008), *Describe Rest Web Services with WSDL 2.0*, http://www.ibm.com/developerworks/webservices/library/ws-restwsdl/ (accessed on November 12, 2012).

Mankin, A. (1991), *RFC-1254: Gateway Congestion Control Survey*, http://tools.ietf.org/rfc/rfc1254.txt (accessed on December 20, 2011).

Marshall, J. (1997), *HTTP Made Really Easy*, http://www.jmarshall.com/easy/http/, last modified: August 15, 1997 (accessed December 12, 2011).

MBTA (2012), *Real time Bus Data*, http://www.mbta.com/rider_tools/developers/default.asp?id=21896 (accessed on September 20, 2012).

McKnight, C., Dillon, A., and Richardson, J. (1991), *Hypertext in Context (Cambridge Series on Electronic Publishing)*, Cambridge University Press.

Meier, J. D. et al. (2009), *Microsoft Application Architecture Guide, Patterns and Practices*, 2nd edition, Microsoft Corporation.

MIC1 (2012), *The National ICT Strategies in Japan are Evolving from e (Electronics) toward u (Ubiquitous)*, http://www.soumu.go.jp/menu_seisaku/ict/u-japan_ en/new_outline01b.html (accessed on October 12, 2012).

MIC2 (2012), *The u-Japan Concept*, http://www.soumu.go.jp/menu_seisaku/ict/ u-japan_en/new_outline03.html (accessed on October 12, 2012).

Microsoft (2009), *Microsoft Application Architecture Guide, 2nd Edition*, http://msdn .microsoft.com/en-us/library/ee658117.aspx (accessed on October 28, 2012).

mobi (2012), .mobi, http://en.wikipedia.org/wiki/.mobi (accessed on September 25, 2012).

Mone, G. (2009), The omnipotence machines, *Scientific American* 301, pp. 58–59.

Mullins, R. (2010), *People Power Releases SDK for Wireless Home Energy Sensors*, http:// venturebeat.com/2010/03/15/people-power-releases-sdk-for-wireless-home-energy-sensors/ (accessed on September 30, 2012).

Nagle, J. (1984), *Congestion Control in IP/TCP Internetworks*, RFC-896, http://tools.ietf .org/html/rfc896 (accessed on December 20, 2011).

NAICS (2012), *North American Industry Classification System*, http://www.census .gov/eos/www/naics/ (accessed on Nov 9, 2012).

National Academy (2007), *Rising Above the Gathering Storm*, National Academies Press.

Nelson, T. H. (1980), Replacing the printed word: a complete literary system, in Lavington, S. H. (ed.), *Proceedings of the IFIP Congress*, Amsterdam: North-Holland, pp. 1013–1023.

Nest (2012), www.nest.com (accessed on September 20, 2012).

Nest Labs (2012), *Nest Learning Thermostat Efficiency Simulation: Update Using Data from First Three Months*; http://www.nest.com/wp-content/themes/nest/static/ white_papers/efficiency-simulation-white-paper.pdf (accessed on August 15, 2012).

NETL (2007), *Modern Grid Benefits*, http://www.netl.doe.gov/smartgrid/referenceshelf/ whitepapers/Modern%20Grid%20Benefits_Final_v1_0.pdf (accessed on September 20, 2012).

Newcomer, E. (2002), *Understanding Web Services: XML, WSDL, SOAP, and UDDI*, Addison-Wesley Professional.

New World Encyclopedia (2012), *Elevator*, http://www.newworldencyclopedia.org/ entry/Elevator (accessed on December 25, 2012).

NFC Forum (2012), *What Is Nfc?* http://www.nfc-forum.org/aboutnfc/ (accessed on December 4, 2012).

NIC (2008), *Disruptive Civil Technologies. Six Technologies with Potential Impacts on US Interests out to 2025*, http://www.fas.org/irp/nic/disruptive.pdf (accessed on October 24, 2012).

Niekamp, R. (2012), *Software Component Architecture*, http://congress.cimne.upc.es/ cfsi/frontal/doc/ppt/11.pdf (accessed on October 3, 2012).

Octopus (2012), http://www.octopus.com.hk/home/en/index.html (accessed on August 7, 2012).

Oki, E., Rojas-Cessa, R., Tatipamula, M., and Vogt, C. (2012), *Advanced Internet Protocols, Services and Applications*, John Wiley & Sons.

Oppenheimer, P. (2011), Top-Down Network Design, third edition, Cisco Press, Indianapolis.

Orlovsky, C. (2005), *Radio Frequency Identification Technology Protects Hospital Patients, Equipment,* http://nursezone.com/nursing-news-events/devices-and-technology/Radio-Frequency-Identification-Technology-Protects-Hospital-Patients-Equipment_24562.aspx (accessed on August 7, 2012).

OSI (2012), *OSI Model,* http://en.wikipedia.org/wiki/OSI_model (accessed on September 22, 2012).

OWL-S (2008), *OWL-S 1.2 Release,* http://www.ai.sri.com/daml/services/owl-s/1.2/ (accessed on November 10, 2012).

Paddison, C. (2004), *Busting the Myths of Pharma RFID,* http://www.atkearney.com/index.php/Publications/busting-the-myths-of-pharma-rfid.html (accessed on August 7, 2012).

Padron, F. M. (2009), *Traffic Congestion Detection Using Vanet,* M.S. Thesis, Florida Atlantic University, Boca Raton, Florida.

Panik, M. J. (2005), *Advanced Statistics from an Elementary Point of View,* Academic Press.

Perry, S. (2008), *Wikitude: Android App with Augmented Reality: Mind Blowing,* http://digital-lifestyles.info/2008/10/23/wikitude-android-app-with-augmented-reality-mind-blowing/ (accessed on December 4, 2012).

Pfister, C. (2011), *Getting Started with the Internet of Things,* O'Reilly Media.

Pogue, D. (2011), *A Thermostat That's Clever, Not Clunky,* http://www.nytimes.com/2011/12/01/technology/personaltech/nest-learning-thermostat-sets-a-standard-david-pogue.html?pagewanted=1 (accessed on August 15, 2012).

Port of Singapore (2012), http://www.singaporepsa.com/aboutus.php (accessed on July 29, 2012).

Price, R. W. (2004), *Roadmap to Entrepreneurial Success,* AMACOM.

Rackley, S. (2007), *Wireless Networking Technology, from Principles to Successful Implementation,* Newnes.

Raghavan, B., McCullough, J., and Snoeren, A. C. (2008), *Embracing Packet Loss in Congestion Control,* http://citeseerx.ist.psu.edu/viewdoc/summary?doi=10.1.1.141.154 (accessed on December 25, 2012).

Rainie, L., and Wellman, B. (2012), *Networked: The New Social Operating System,* MIT Press.

Rellermeyer, J. S. et al. (2008), *The Software Fabric for the Internet of Things,* in IOT 2008, Floerkemeier, C. et al. (eds), LNCS 4952, pp. 87–104.

RFC 791 (1981), *Internet Protocol,* http://tools.ietf.org/html/rfc791#section-3.1 (accessed on September 13, 2013).

RFC 793 (1981), *Transmission Control Protocol,* http://tools.ietf.org/pdf/rfc793.pdf (accessed on January 1, 2013).

RFC 2396 (1998), *Uniform Resource Identifiers (URI): Generic Syntax,* http://tools.ietf.org/html/rfc2396 (accessed on September 30, 2012).

RFC 3986 (2005), *Uniform Resource Identifier (URI): Generic Syntax,* http://tools.ietf.org/html/rfc3986 (accessed on September 30, 2012).

RFID 2012, *How Much Does an RFID Tag Cost Today?* http://www.rfidjournal.com/faq/20/85 (accessed October 17, 2012).

RFID Update (2007), *RFID Solution Tracks 100,000 Individual Documents,* http://www.rfidjournal.com/article/view/6869 (accessed on December 31, 2012).

RIR (2013), https://www.iana.org/numbers (accessed on September 13, 2013).

Roberti, M. (2004), *The History of RFID Technology,* http://www.rfidjournal.com/article/articleview/1338/1/129 (accessed on January 14, 2012).

Roberti, M. (2006), *The Lahey Clinic's RFID Remedy,* http://www.rfidjournal.com/article/view/2265 (accessed on July 30, 2012).

Roberti, M. (2010a), *Wal-Mart Relaunches EPC RFID Effort Starting with Men's Jeans and Basics,* http://www.rfidjournal.com/article/view/7753 (accessed on January 5, 2011).

Roberti, M. (2010b), *The Internet of Things Revisited,* http://www.rfidjournal.com/article/print/7638 (accessed on January 2, 2013).

Roberti, M. (2011a), *Medicarte Uses RFID and Biometrics to Reduce Counterfeiting,* http://www.rfidjournal.com/article/view/9065 (accessed on August 7, 2012).

Roberti, M. (2011b), *Retailers Say Shrinkage Rose in 2011,* http://www.rfidjournal.com/blog/entry/8899/ (accessed on August 6, 2012).

Roberti, M. (2012), *Reducing Health-care Errors,* http://www.rfidjournal.com/blog/entry/9774/ (accessed on August 7, 2012).

Roduner, C. (2010), *BIT—A Browser for the Internet of Things,* http://www.vs.inf.ethz.ch/publ/papers/roduner-bit-2010.pdf (accessed on October 8, 2012).

Ruhanen, A. et al. (2008), *Sensor-enabled RFID Tag Handbook,* http://www.bridge-project.eu/data/File/BRIDGE_WP01_RFID_tag_handbook.pdf (accessed on January 17, 2012).

Samsung (2012), *Connected Together,* http://www.samsung.com/us/connected-devices/ (accessed on December 4, 2012).

Sarma, S., Brock, D., and Ashton, K. (2000), *The Networked Physical World, Proposals for Engineering the Next Generation of Computing, Commerce and Automatic-Identification,* http://www.autoidlabs.org/uploads/media/MIT-AUTOID-WH-001.pdf (accessed on March 27, 2013).

Scheithauer, G., and Winkler, M. (2008), *A Service Description Framework for Service Ecosystems,* Issue 78 of Bamberger Beiträgezur Wirtschafts informatikundangewandten Informatik, Fak. Wirtschaftsinformatik und Angewandte Informatik, Otto-Friedrich-Univ.

Schuster, E. W., Allen, S. J., and Brock, D. L. (2007), Global RFID, *The Value of EPCglobal Network for Supply Chain Management,* Springer.

Sen.se (2012), http://open.sen.se/ (accessed on October 15, 2012).

SentryGPSid (2009), *Alarming Lost Child Statistics,* http://sentrygpsid.com/GPS/child-gps/alarming-lost-child-statistics (accessed on July 29, 2012).

Shoup, D. (2005), *The High Cost of Free Parking,* APA Planners Press.

Siegel, J. (2000), *CORBA 3, Fundamentals and Programming,* John Wiley & Sons.

SIFMA (2010), *US Financial Services Industry,* http://www.ita.doc.gov/td/finance/publications/U.S.%20Financial%20Services%20Industry.pdf (accessed on November 2, 2012).

SmartGrid (2012), http://www.smartgrid.gov/the_smart_grid#smart_grid (accessed on September 20, 2012).

Spies, B. (2008), *Web Services, Part 1: SOAP versus REST,* http://www.ajaxonomy.com/2008/xml/web-services-part-1-soap-vs-rest (accessed on November 12, 2012).

Spohrer, J., Maglio, P. P., Bailey, J., and Gruhl, D. (2007), Steps toward a science of service systems, *Computer,* Vol. 40, Issue 1, pp. 71–77.

Srinivasan, A. (2005), Biographies of *Modern Inventors,* Sura Books, Madras, India.

Srisuresh, P. and Holdrege, M. (1999), *IP Network Address Translator (NAT) Terminology and Considerations,* RFC 2663, http://tools.ietf.org/html/rfc2663 (accessed on September 13, 2013).

Steinmetz, R., and Wehrle, K. (2005), *Peer-to-peer Systems and Applications,* Springer.

Streetline (2012), http://www.streetline.com/parksight/ (accessed on September 20, 2012).

Sullivan, L. (2009), *How RFID will help Mommy Find Johnny,* http://www.information-week.com/news/47208448 (accessed on July 29, 2012).

Sundmaeker, H., Guillemin, P., Friess, P., and Woelffle, S. (2010), *Vision and Challenges for Realizing the Internet of Things,* European Commission—Information Society and Media DG.

Swedberg, C. (2009), *Hong Kong Airport Says It Now Uses only RFID Baggage Tags,* http://www.rfidjournal.com/article/print/4885 (accessed on July 30, 2012).

Swedberg, C. (2010), *ORLocate RFID-enabled System for Surgical Sponges and Instruments gets FDA Clearance,* http://www.rfidjournal.com/article/view/7836 (accessed on July 30, 2012).

Swedberg, C. (2012), *Tread for London Bus Company,* http://www.rfidjournal.com/article/view/9697/1 (accessed on August 7, 2012).

Sweeney, P. J. (2005), *RFID for Dummies,* Wiley Publishing Inc., Indianapolis.

Thien, H. P., Moelyadi, M. A., and Muhammad, H. (2008), *Effects of Leader's Position and Shape on Aerodynamic Performances of V Flight Formation,* http://arxiv.org/ftp/arxiv/papers/0804/0804.3879.pdf (accessed on August 22, 2012).

ThingWorx (2012), http://www.thingworx.com/ (accessed on October 15, 2012).

3M (2012), *Case Studies and Whitepapers,* http://solutions.3m.com/wps/portal/3M/en_US/rfid_tracking_solutions/home/resources/case_studies/ (accessed on December 31, 2012).

TLDs (2012), http://data.iana.org/TLD/tlds-alpha-by-domain.txt (accessed on February 15, 2012).

Trading Economics (2011), *Personal Computers (per 100 People) in the United States,* http://www.tradingeconomics.com/united-states/personal-computers-per-100-people-wb-data.html (accessed on December 17, 2011).

Ubiquitous ID Center (2006), *Ubiquitous ID Architecture,* http://www.uidcenter.org/wp-content/themes/wp.vicuna/pdf/UID-CO00002-0.00.24_en.pdf (accessed on October 13, 2012).

uCode (2012), *Usage Example,* http://www.uidcenter.org/usage-example/usage-example (accessed on October 13, 2012).

UDDI (2006), *Microsoft, IBM, SAP to Discontinue UDDI Web Services Registry Effort,* http://soa.sys-con.com/node/164624 (accessed on November 11, 2012).

uID (2012), http://www.uidcenter.org/ (accessed on October 13, 2012).

uID (2013), http://www.uidcenter.org/usage-example/usage-example#guide (accessed on March 27, 2013).

UNSPSC (2012), *United Nations Standard Products and Services Code,* http://www.unspsc.org/ (accessed on November 4, 2012).

UPC Code, 2008, *UPC Code Sample,* https://www.upccode.net/upc_sample.html (accessed on January 14, 2012).

US Census (2011), *Computer and Internet Use in the United States: 2011,* http://www.census.gov/hhes/computer/publications/2011.html (accessed on October 1, 2013).

US Census (2012), *Current Population Clock,* http://www.census.gov/main/www/popclock.html (accessed on December 4, 2012).

Vargas, J. A. (2012), *Spring Awakening: How an Egyptian Revolution Began on Facebook*, http://www.nytimes.com/2012/02/19/books/review/how-an-egyptian-revolution-began-on-facebook.html (accessed on March 13, 2013).

Vasseur, J.-P., and Dunkels, A. (2010), *Interconnecting Smart Objects with IP*, Elsevier.

Vestberg, H. (2010), *CEO to Shareholders: 50 Billion Connections 2020*, http://www.ericsson.com/news/1403231 (accessed on January 5, 2012).

Violino, B. (2005), *At Anglo American Platinum's Paardekraal Mine in South Africa*, RFID is Saving Dollars—and Lives, http:www.rfidjournal.com/article/view/1759 (accessed July 30, 2012).

Violino, B. (2010), *Memorial Hospital Miramar Builds Benefits onto its RTLS*, http://www.rfidjournal.com/article/purchase/7431 (accessed on December 31, 2012).

W3C (2004), *RDF Primer*, http://www.w3.org/TR/rdf-primer/#ref-rdf-concepts (accessed on October 31, 2012).

W3C (2012), *The Web and Mobile Devices*, http://www.w3.org/Mobile/ (accessed on September 25, 2012).

W3C (1999), *A Little History of the World Wide Web*, http://www.w3.org/History.html (accessed on February 17, 2013).

W3C-Schema (2012), *Schema*, http://www.w3.org/standards/xml/schema (accessed on September 12, 2012).

W3Schools (2012), *WSDL Tutorial*, http://www.w3schools.com/wsdl/default.asp (accessed on November 11, 2012).

Waldner, J.-B. (2010), *Nanocomputers and Swarm Intelligence*, John Wiley & Sons.

Wallop, H. (2011), Japan Earthquake: How Twitter and Facebook Helped, http://www.telegraph.co.uk/technology/twitter/8379101/Japan-earthquake-how-Twitter-and-Facebook-helped.html (accessed on March 13, 2013).

Weber, R. H., and Burri, M. (2013), *Classification of Services in the Digital Economy*, Springer, Berlin, Germany.

Weiser, M. (1991), The computer for the 21st century, *Scientific American*, Vol. 265, Issue 3, pp. 94–103.

Wessel, R. (2009), *Lisbon Airport Ups Throughput with RFID Baggage System*, http://www.rfidjournal.com/article/view/5302 (accessed on January 5, 2011).

Williams, R. (2008), *What is the Real Business Case for the 'Internet of Things'?*, http://www.itsc.org.sg/pdf/synthesis08/Five_Internet.pdf (accessed on October 17, 2012).

Winett, J. M. (1971), *The Definition of a Socket*, http://tools.ietf.org/html/rfc147 (accessed on March 27, 2013).

Wolf, H.-J. (1974), *Geschichte de Druckpressen*, 1st ed., Frankfurt/Main: Interprint.

World Bank (2011a), *Internet Users (per 100 People)*, http://data.worldbank.org/indicator/IT.NET.USER.P2/countries/1W?display=graph (accessed on December 14, 2011).

World Bank (2011b), *Mobile Cellular Subscriptions (per 100 People)*, http://data.worldbank.org/indicator/IT.CEL.SETS.P2/countries?display=graph (accessed on December 16, 2011).

World Bank (2011c), http://data.worldbank.org/ (accessed on October 17, 2012).

Xfinity (2012), http://www.comcast.com/homesecurity/security.htm (accessed on August 15, 2012).

Yahoo (2012), http://pipes.yahoo.com/pipes/ (accessed on October 28, 2012).

Yan, H. (2011), *China sets 5b Yuan Fund for IOT Industry*, http://www.chinadaily.com.cn/bizchina/2011-08/23/content_13172322.htm (accessed on September 30, 2012).

Yeager, N. J., and McGrath, R. E. (1996), *Web Server Technology, The Advanced Guide for World Wide Web Information Providers*, Morgan Kaufmann.

Youtube (2012b), http://www.youtube.com/t/press_timeline (accessed on October 16, 2012).

Zhao, Z., Laga, N., and Crespi, N. (2009), *A Survey of User Generate Service*, IEEE International Conference on Network Infrastructure and Digital Content, IC-NDC, Evry, France.

Zhou, J. H., Ni, L., and Mutka, M. (2003), *Prophet Address Allocation for Large Scale MANETs*, Proc. of IEEE INFOCOM.

Index

For Product Safety Concerns and Information please contact
our EU representative GPSR@taylorandfrancis.com Taylor & Francis
Verlag GmbH, Kaufingerstraße 24, 80331 München, Germany

T - #0163 - 230425 - C248 - 234/156/11 - PB - 9780367379186 - Gloss Lamination